SAWDUST IN THE WESTERN WOODS

George Youst sawmill set up in the woods, ten miles from Allegany, Oregon – above the Golden and Silver Falls, 1940.

SAWDUST IN THE WESTERN WOODS

A personal, pictorial, and primarily oral history of the "gyppo sawmill" in the Douglas fir region, 1926-1956

by

Lionel Youst

From taperecorded interviews conducted in 1967 with his father, George Youst, a small sawmill operator for practically all the years that the phenomenon existed, supplemented by personal experience of the author and supported by various archival and other sources.

Golden Falls Publishing   **Allegany, Oregon**

Golden Falls Publishing, Allegany, Oregon ©
2009 by Lionel Youst
        second impression, September, 2009
        third impression, July, 2020
Printed in the United States of America ISBN
0-972622 6-2-4

Library of Congress cataloging data: HD9757
Dewey Decimal cataloging data: 338.47674

        Youst, Lionel D., 1934 -

                Sawdust in the Western Woods
                includes index, maps, bibliography, and photographs

                Lumber trade – Washington – Cowlitz County – History
                Lumber trade – Oregon – Coos County – History
                Cowlitz County, Washington – History
                Coos County, Oregon – History

Epigraph, page v, is from Robert E. Swanson's poem "The Kickback of Fate" as it appeared in *Rhymes of a Western Logger* (1942, 1992), used by permission of the publisher, Harbour Publishing, Box 219, Madeira Park, BC VON 2HO, Canada. www.harbourpublishing.com

Part of chapter 3 (The Gyppo Sawmill, Coos County) of the present work appeared in a slightly different form in *Above the Falls: an oral and folk history of Upper Glenn Creek, Coos County, Oregon*, [1992] 2nd ed. 2003.

All photographs are from the author's family albums except the photos on page 21 and 25, which are from The Timberman, April and May, 1934. Advertisements are from pre-1934 editions of The Timberman, copies of which are in the author's possession.

The first impression was printed in 100 copies on the occasion of the "first annual" Forest History Roundtable Conference held at Tillamook Forest Center, Tillamook, Oregon, May 15, 2009.

To contact the publisher:

Lionel Youst
12445 Hwy 241
Coos Bay, OR 97420
(541) 267-3762
lionel@youst.com

So back to our work on the "edger," for the head-rig was screaming again;
Where the "carriage" receded like thunder and the sawdust it pattered like rain.
The "cants" fell down on the "roll-case," with the boom of a Congo drum,
While the sawyer conversed with the "setter" in signs of the deaf and the dumb.

Did you ever stand in a sawmill and hear those unearthly sounds,
When the trimsaws barked and they jangled like the yapping of quarrelling hounds --
The dual-tone cry of the edger, distinctive of sawmilling lore,
While the timber-sizer continues with its "diapasonal" roar?

Well, a man in a mill only hears them when something is eating his heart,
And I heard them quite plainly that morning, as my helper was playing his part.

– Robert E. Swanson
"The Kickback of Fate"

Contents

One	An Essay on "Gyppo"	1
Two	The Tie Mill, Cowlitz County, Washington: 1926-1937 an oral history	13
Three	The Gyppo Sawmill, Coos County, Oregon: 1937-1955 an oral history (continued)	45
Four	An Epilogue: How it all Ended	78
Maps	Cowlitz County, Youst tie-mills Coos County, Youst gyppo-mills	12 44
Glossary		89
Bibliography		90
Index		92

To the Reader

George Youst, my dad, owned and operated small sawmills in Washington and Oregon from 1927 until 1956. There are hundreds of books that cover aspects of logging during that period, but none about the small sawmills that were set up in the woods with the logging done directly to them. As far as I know, this is the first published account of the evolution and metamorphosis of those kinds of mills. My purpose is to make available what I know about it, based on extensive taperecorded interviews I did with my dad in November, 1967, and supplemented by my own experience and other sources. Pictures included in this volume are mostly from my family's picture album.

There may be a certain amount of duplication. The original draft of the essay on "Gyppo" was intended for publication with the Oregon Historical Quarterly but was rejected on the basis of irrelevance. It is hoped that a more selective readership will find it of some use. The rest of the volume consists of the lightly edited and extensively annotated transcripts of the 1967 interviews and are offered as primary source material for anyone who wishes to use it. Chapter 2 – Cowlitz County – had been submitted to the Cowlitz County Historical Quarterly but was undergoing more extensive editing than I was willing to put up with. Parts of Chapter 3 – Coos County – appeared in somewhat different form in my book, *Above the Falls: an oral and folk history of Upper Glenn Creek, Coos County, Oregon*, [1992] 2nd edition 2003. The Epilogue is all new material, written for this volume.

The oral interviews that appear herewith are documents presented as a case history of the small sawmills which first emerged during the 1920's, using the internal combustion gasoline engine as a reliable power source. They were set up in the timber with logging done directly to the mill, and they primarily cut railroad cross-ties until the late 1930's when increasingly they began to also take orders for dimensioned lumber. During World War II the small sawmill came into its own and by the end of the war they were producing about one-quarter of all the lumber produced in the United States. After the end of the war the small sawmills proliferated, especially in the Douglas fir regions of the far west. Those post-war small mills came to be known as "Gyppo Sawmills," and the small sawmills of the previous period were retrospectively given the same label. It is one of the objects of this volume to set the record straight on that score!

Logging history has hundreds of volumes of books. Sawmilling has almost none. This is probably the first book to focus on the sawmiller who was a logger adding value to his logs by manufacturing them into lumber, in situ.

GYPPO
selected historical uses of the term

gyppo, gypo, n. **gyppoing**, v.
probably from Gypsy + o.

1. **gyppoing**, verb.

   1919. *Industrial Worker* [Seattle], editorial, 'The Gypsy's Warning.' "At present the master class of capitalists call it 'contract labor,' piece work,' and other fancy names . . . . For us, the proletarians, it is 'gyppoing' and it means all that the name connotes.[1]

2. **gyppo, gypo**, noun. Contract or piece labor, especially in logging and sawmilling..

   1919: *Industrial Worker* [Seattle], editorial, "The Gypsy's Warning." "The gyppo is a man who 'gyps' his fellow workers and finally himself, out of the fruits of all our organized victories in the class war."[2]

   1920: Ralph Winstead, "Enter a Logger: An I. W. W. Reply to the 4-L's," *Survey*. 44 [July 3, 1920], p. 475. "Now it's the same way with the Gypo proposition. These here bushel maniacs just naturally got the same sort of nature as John D. Rockefeller and a lot of other humans includin' all of us."

   Ibid. "'Wanted eight men to take contract bucking and falling. Details inside.' There it was, straight Gypo stuff. Chance to make a co-operative fortune right in my hand." [3]

   1922: *Industrial Worker*, October 7, 1922. "The gyppo system was introduced in the logging camps universally, and in a general way in the fall of the year 1919."[4]

   1923: E. B. Mittelman, "The Gyppo System," *The Journal of Political Economy*, v. 31, no. 6 [Dec. 1923]. "It is not always easy to distinguish a contractor from a gyppo."

   1937: Mrs. Edna C. Dawson, poem "Hail the Gypo!"[5]

   1938: Steward H. Holbrook, *Holy Old Macinaw: A Natural History of the American Lumberjack.* "Gyppo: Any sort of contract work. See By the Bushel."

   Ibid. "By the Bushel, by the inch, by the mile: contract work, but usually applied to contract falling and bucking."

1941: Robert J. Filberg, manager, Comox Logging and Railway Co., Ltd. British Columbia, address to the 32nd Logging Congress in Seattle, Washington, October, 1941.[6] "But let us remember it was not the big or high-ball logging operators nor the mechanical or logging engineer, who conceived, pioneered and developed truck logging. It was the Gyppo, and ranking with Paul and Babe, in the loggers' niche of fame is a Gyppo and his truck. Gentlemen, I propose a toast to our friend the Gyppo."

1952: Margaret Elley Felt. "I'm a Gyppo Logger's Wife," *Saturday Evening Post*, August 30, 1952.

1953: Steward H. Holbrook. *The Timberman*, September, 1953, p. 5. "and he seems to thoroughly enjoy being a salvage operator – independent, not gyppo."

1964: Ken Kesey. *Sometimes a Great Notion*, p. 599. "Scabbin'! For a goddam gyppo outfit who wouldn't meet the union regs!"

1988. William G. Robbins. *Hard Times in Paradise: Coos Bay, Oregon, 1850-1986*. p. 110. "The immediate postwar years in southwestern Oregon were the heyday of the storied gyppo logger and sawmill operator, the hardy individual who worked on marginal capital, usually through sub-contracts with a major company or a broker . . . Reckless and daring gamblers, always ready to move on to the next stand of timber, the multitudes of gyppos were unique to the postwar era in the Douglas fir country."

Ibid, pp. 110-11. "George Youst, who came to the bay area from southwestern Washington, where he milled railroad ties and bridge planks, was one of the earliest of the modern-day gyppos in Coos County."

An essay on "Gyppo Sawmills"

Introduction

"Gyppo" figures prominently in the twentieth-century economic and social history of Oregon. Immortalized by the fictional Stamper family of Ken Kesey's novel, *Sometimes a Great Notion,* it has come to denote a distinct subculture in the Pacific Northwest. Coined during a labor strike at the time of World War I, it was a pejorative term for contract labor in logging camps and sawmills. By the 1930's it had radiated to include owner-operated logging trucks and capital-intensive logging contractors. After World War II it included the numerous small sawmills that sprung up almost anywhere that 80 or 160 acres of suitable timber stood in non-industrial private ownership. Those mills were the natural successor to the semi-portable, gasoline engine driven, tie-mills that had proliferated during the 1920's. Economically, "gyppo" suggests an example of the hypothetical construct known as "perfect competition." Linguistically, it provides an interesting example of amelioration, in which the meaning of a word becomes more and more favorable, or positive, over time. If it had appeared in the Oxford English Dictionary its entry might have taken the form used in the preceding two pages.

A Gyppo Case in Point

In 1967 I recorded nine hours of interviews with my dad, George Youst. He had been what came to be known as a "gyppo" sawmiller during all the years that there was such a thing, and those interviews give a pretty good case study in the evolution of the gyppo phenomenon. He was born in 1899 and worked as a logger from the time he was 14. In 1926 he and a partner had a contract to load logs out of the Chehalis River onto railcars at his hometown of Centralia, Washington. This was the classic "gyppo." All of the equipment belonged to the company and there was no capital investment on his part. He relied on his own skill and knowledge, and hired extra hands as needed. The contract was to load the logs, and how it was done was up to him.

The owner of the timber, after observing the operation through the course of the summer, told him, "Fellows that work like you guys do could make some money in a tie mill." My dad didn't know what a tie mill was, but he checked into it and found that standing timber suitable for manufacture into railroad ties was going at $1.50 per thousand board feet. The going price for ties was $18.25 per thousand, delivered and loaded into box cars. It appeared that there was enough margin for a profit, even for someone who knew nothing about sawmills. He located a suitable 160 acres of timber above the highway between Kalama and Woodland, Washington, asked his uncle to co-sign a note for him at the bank, and sold his Model-T Ford car. With that he and a partner made down payments on a used, portable sawmill powered by a Fordson tractor engine and a used Chevrolet flatbed truck to haul the ties on. His logging was "gyppoed out" to a farmer with a team of horses, until he made enough money to buy his own team. The men who fell and bucked the timber were paid by the tie. The idea of "getting gyped" came up early in the operation. His partner took his turn as head sawyer for awhile, cutting a boxcar load of ties. The railroad inspector culled 20% of them for "mis-manufacture," which meant that the crew got 20% less than what should have been due them on that carload. These were the hazards of working as a "gyppo" in those days.

When I was born, in January, 1934, we lived in a board and batten shack a couple hundred feet from my dad's tie mill on the Little Kalama River in Cowlitz County, Washington. We were in the center of a half-section (320 acres) of mostly Douglas fir timber he bought cheap in 1930 from the Acme Coat Hanger Company of Vancouver, Washington, which had originally purchased the land for the alder, used in wooden coat hangers.

In those days small mills that cut mostly railroad ties were usually set up in the timber, which was logged directly to the mill. They were called tie mills, not gyppo mills. The term "gyppo" was applied to them retrospectively, after World War II. In southwestern Washington in the early 1930's there were about 125 of these tie mills trying to eke out a living from a railroad industry that was replacing fewer railroad ties than at any time in living memory.

Fairhurst Lumber Company of Tacoma was the brokerage firm that tried to apportion orders to all of those mills, but orders from the railroads were small and they were infrequent. Fairhurst would have needed to receive an order for 125 carloads of ties before they could apportion out even one carload (25,000 board feet) for each sawmill, and thus the mills all had to wait their turn, and sometimes the wait was long. After the bottom fell out of the market in 1931, the sawmills received as little as $6.82 per thousand for the ties, down from a high of $18.25.[7] Those tiemills that survived depended on occasional orders for bridge plank or timbers from the county road departments, or miscellaneous lumber for nearby construction projects. The times were very hard indeed.

By 1937 the economy was starting to pick up somewhat and that year Herbert Busterud, an official of the Coos Bay Lumber Company, formed his own brokerage firm in hopes of attracting a number of small sawmills into southwestern Oregon.[8] At Busterud's encouragement, Dad moved his mill from Washington to Oregon in early 1937 and set it up on an 160-acre patch of "third-growth" Douglas fir a few miles from North Bend. It was among the earlier of the uncounted hundreds of small sawmills that were eventually set up to operate in Oregon during the years following the end of World War II.

A New Power Source

It was the internal combustion engine – first gasoline, and later diesel – that made these semi-portable sawmills possible. Before the advent of a practical gasoline engine, sawmills relied first on water power, later on steam. Neither of those two traditional power sources lent themselves to mobility. Those early mills were semi-permanent facilities and logs had to be brought to them by rail or water. The widespread use of internal combustion engines changed that, and permitted the sawmills to be disassembled and moved as the timber around them ran out and new sources of timber were secured. A typical setting might last one or two years, sometimes less, but seldom longer. Depending on the degree of mobility built into the mill, it could take anything from a few days to a month or more to move the mill and get it running again at a new setting, but it was something that was quite doable because the power source could be easily moved and was not site-dependent.

The sawmill my dad built in 1940 (the earlier one burned in 1938) came at the time that General Motors was first promoting its new stationary two-cycle diesel engine, later known as the Detroit Diesel.[9] These engines were advertised as using only 25 gallons of diesel to run a sawmill cutting 15,000 board feet in eight hours. Diesel was only about seven cents per gallon at that time, and the mill could thus run for eight hours on $1.75 worth of fuel.[10] In 1940, Dad bought a six-cylinder GM diesel engine for $3,660 from the dealer, Gunderson Brothers of Portland, paying $700 down and $87.22 per month until it was paid off.[11] This lifted the operation to a new level. With the dependable 160 hp that the new engine provided, an edger was installed and orders were taken for lumber of almost any dimension. Later, a second GM diesel gave the mill one engine to power the headrig and feedworks, and the second engine to power the edger, cutoff saws, and conveyors. This made a very efficient and powerful sawmill that produced 20 to 24 thousand board feet of dimensioned 24-foot lumber per eight-hour day, a fairly good output for mills of its class at that time.

Small sawmills in general

The small sawmills of the period prior to the end of WW II, those cutting less than 50,000 board feet per day, could be divided into three classes.[12] The smallest of the mills could be considered family or neighborhood sawmills, and they were set in a permanent location, usually on somebody's farm. Logs were usually brought to them by truck and they cut less than 5,000 board feet per day, and not every day. For the most part they were owned and operated by farmers, who ran them only when they had local orders received directly from their neighbors.

The second class of sawmill would be the "tie mills," almost all of which were set up in the timber. The timber was logged directly to the mill, frequently by high-lead or skyline yarders, sometimes by Cat logging, or both. When all the timber surrounding the mill for perhaps 2500 feet or so had been cut off, the mill had to be moved to a new stand of timber or cease operation altogether. The tie mills typically cut 10 to 25 thousand board feet per day and marketed their railroad ties through a broker. These mills typically cut eight-foot logs and they sometimes doubled as a "stud mill," cutting eight-foot two-by-fours from the side-cut of the ties.

The third class of small sawmill of that period was set up to manufacture dimensioned lumber. It differed from the others in that it had more power, and it had an edger that allowed it to produce a wide variety of dimensioned lumber, in addition to planks, timbers, and ties. Most of these mills could produce from 20 to 50 thousand board feet per day of lumber up to 24 feet long. A few of them even had a planer to finish the lumber. Many of them were set up in the timber, in the manner of the smaller tie mills but more often logs would be hauled in by truck from a logging operation that was nearby. If it was more than a few miles to the logging, the mill would usually be moved to be closer to its timber supply. During World War II this class of sawmill grew in numbers because of the insatiable demand for lumber for the war effort. By the end of the war, they were putting out more than 25% of the total lumber in the United States.[13]

There had been governmentally enforced wage and price controls on lumber throughout the war but those restrictions were lifted in November, 1946.[14] Fueled by the post-war housing boom and a lively international market, the price and demand for lumber increased rapidly at the same time that component equipment for small sawmills was becoming more readily available from many manufacturers. There was enough merchantable timber available in non-industrial private ownership to make the scramble to get it seem worthwhile. This lasted for a few short years, from 1947 until about 1957, and the proliferating small sawmills of that period are the ones that we usually think of when we talk of "gyppo mills."

The Term Ameliorates (becomes better!)

The end of the war changed everything. The gasoline chainsaw, for example, only came into general use after the end of the war. Timber fallers had been paid by their output in most logging camps for at least 25 years. They were paid so much per thousand board feet of trees fell or logs bucked. They were known as "bushlers," paid "by the bushel," to use the analogy of agricultural pickers. At the end of the war, some of them bought the new gasoline chain saws for around $400 and, being paid at the same rate as they had been getting for hand sawing, they sometimes made ten times as much. They proudly called themselves "gyppos" and were the envy of many wage-earning union loggers.[15] $100 per day was the goal, and many made it, at a time when $12 per day was considered good money in the woods. It was also about this time that the small sawmills, both tie mills and the larger mills that produced dimensioned lumber, began universally to be called Gyppo mills, not as a pejorative term, but merely as a descriptive one.

My mother thought the term derived from "gypsy" because, she said, the small mills moved so much, like gypsies.[16] At the end of the war, I was twelve years old and in my short life we had moved seven times, following the timber with a small sawmill. Her etymology made sense to me. My dad, however, did not use the word. He remembered the days of the I.W.W. and the manner in which the "gyppo system" was introduced in an attempt to break the union back in 1919. There was the lingering sense of something unsavory, something perhaps unethical or even illegal hovering about the term. In 1953 Stewart H. Holbrook wrote a personality sketch of Frank Donahue, a gyppo logger of Port Angeles, Washington. "Frank dislikes the term 'gyppo,' as applied to contract loggers," Holbrook explained. "In fact, he gets rather hot about it, and will argue if need be for hours to convince the ignorant that gyppo has connotations of crooked dealing. 'Independent logger' is the right term." Between 1948 and 1953, Holbrook wrote 48 sketches of logging personalities and only two of them contain the word "gyppo."[17]

Today, in the early years of the 21st century, virtually all logging in the northwest states is done by independent contract loggers. They are all known as "gyppos," and there is no negative connotation remaining in the term. The gyppo sawmill has disappeared from the economy and almost from memory, and the big mills have outsourced their logging to gyppo loggers, who accept the term with pride.

Gyppo Economics

The economic conditions that gave rise to the "Gyppo sawmill" phenomenon was fleeting and bears a certain resemblance to the hypothetical market form known to economists as "perfect competition."[18] This is not to say that competition in the sawmill business was ever perfect, far from it. But the rapid increase in the number of small sawmills in Oregon during the ten years following the end of World War II suggests an economic environment that had some of the elements of "perfect competition." Examples usually given concern agriculture commodities. Potatoes, for example, are more or less all alike and they are grown by many farmers and purchased by many consumers, none of whom have enough market power to influence the price.

The boom in gyppo sawmills (at least in the Douglas fir region) had some similarities to the example of the potatoes. The sawmill product, dimensioned lumber, was pretty much the same no matter which mill it came from. There were hundreds of sawmills, and hundreds of purchasers, none of whom could by themselves affect the price, which under the circumstances was determined by the law of supply and demand. There are, however, a couple other requirements that must be fulfilled to meet the criteria of "perfect competition." Everybody has to have the same information as to what the prices are. In general, this criteria was met. Everybody knew what the price of ties or two-by-fours were. Also, there has to be a certain ease getting into the business. For those few with an aptitude and ability to run a sawmill, the only real hurdle was locating a supply of timber that would provide the leverage needed to borrow enough money for a mill. But it all fell apart in the end. The stumpage price of standing timber went up and up as the supply of timber available to the excessively large number of small operators decreased precipitously. By the mid 1950's the boom was over, but for the timber towns of the Pacific Northwest, it was a great ride while it lasted.

ENDNOTES

1. Quoted in E. B. Mittelman. "The Gyppo System." In *The Journal of Political Economy*, vol. 31, no 6 [December, 1923]. The University of Chicago Press. p. 840.

2. Ibid.

3. Quoted in Joyce L. Kornbluh, ed. *Rebel Voices: An I. W. W. Anthology.* University of Michigan Press, 1964, 1968. p. 282.

4. quoted in E. B. Mittelman. "The Gyppo System." In *The Journal of Political Economy*, vol. 31, no 6 [December, 1923]. The University of Chicago Press. p. 850.

5. The poem, "Hail the Gypo!" by Mrs. Edna C. Dawson, appeared in *The Timberman*, September, 1937, p. 72.

6. Reprinted as "Lessons from a Gyppo Logger" in *The Timberman*, November, 1941, pp. 30-2.

7. The details of tie mill operations in southwestern Washington during the Depression were obtained from tape-recorded interviews with George Youst during September, 1967.

8. The following articles from *The Timberman* are indicative of the increase in small sawmills prior to World War II, facilitated by the role of a brokerage firm. These small tie mills were precursors to the many small mills that appeared during and after World War II, and came to be known generally as "gyppo mills." *The Timberman*, July 1937, p. 82. "Marshfield, Ore, July 15. Busterud Lumber Co., a new Marshfield corporation, has established a tie terminal here for domestic and export handling of ties, which will be cut in a number of small mills throughout Coos County." *The Timberman*, September 1937, p. 94: "Busterud Lumber Co., H. A. Busterud manager, Marshfield, conducting a domestic and export tie business, is operating 14 tie mills, producing about 150,000 feet of ties and lumber daily." The *Timberman*, August 1940, p. 30: "An assembling and shipping service for 26 small Coos County sawmills is maintained at Marshfield by H. A. Busterud, operating as the Busterud Lumber Co." Later, Busterud's operation was purchased by the older and larger brokerage firm, Fairhurst Lumber Co. *The Timberman*, March 1941, p. 76: "Fairhurst Lumber Co. Of Tacoma, Washington, has leased Portland Docks Co.'s dock at Marshfield, Oregon, and will continue operation on the former basis conducted by H. A. Busterud."

9. *The Timberman*, July, 1940, p. 67: "George Yust (sic) is building a 10,000 capacity sawmill at Allegany, operated by a diesel engine. Lumber will be trucked 30 miles to Coos Bay."

10. *The Timberman*, April 1940, has the first of the many advertisements showing testimonials of sawmill operators who used the GM diesel engine. The one purchased by my dad was serial number 284. Four years later, the serial numbers ran well above 15,000, attesting to the popularity of the engine.

11. Conditional Sales Contract, July 26, 1940: George Youst, purchaser; Gunderson Brothers of Portland, Oregon, seller. George Youst sawmill file, in the author's possession.

12. Herman M. Johnson. "The Small Sawmill in the Douglas Fir Region," *The Timberman*, December 1944, pp. 34-6. This article is a very good analytical discussion of the small sawmills west of the Cascade Mountains of Washington and Oregon during the 1940's.

13. Ibid. A chart in the article, using data from the U. S. Forest Service and the Bureau of Census, shows that in 1929 there were 383 small sawmills in Oregon. That is, those cutting less than 50 thousand board feet per day, the usual designation of a small sawmill. Those 383 sawmills produced 17.43% of the total lumber produced in the state that year. In 1943 the number of small sawmills had increased to 502, and their share of total lumber production was now 27.29%. This increase was primarily due to the wartime demand for lumber, a demand the large lumber companies could not meet by themselves. The wartime increase in the number of small sawmills gave the impetus for the astonishing increase in the number of small sawmills in the years immediately following the end of the war. All data concerning small sawmills must be used with caution, however. Those that were under contract to fill orders from a large mill frequently were not included in the statistics. The total output in those cases was usually credited to the large mill. Many small sawmills were lost from the statistics in that way and it may thus be assumed that the actual production from small sawmills was significantly greater than the record shows.

14. The Timberman, May 1947, p. 33.

15. *The Timberman* was slow to begin using the term "gyppo." Perhaps the first was a poem, "Hail the Gypo," by Mrs. Edna C. Dawson, published September 1937, page 72. "In the distance I see a picture with a diesel coming through / Come courageous gypos, your dreams are coming true." With the exception of that poem and the article by Robert J. Filberg, "Lessons from a Gyppo Logger," November 1941, pp. 30-2, *The Timberman* avoided use of the term "Gyppo" until well after the end of World War II. The editor and publisher, George M. Cornwall (1867-1950), no doubt understood the historically negative connotations of the word and refrained from using it. In 1948 advertisements began to appear in the journal for a gasoline falling saw called the "Woodsman Gypo," and after that the term showed up in a number of articles. The term was becoming respectable.

16. The role of the wives of gyppo sawmill and logging operators has been a sadly neglected field of study. My mother, for example, was a full partner in the operation. She kept the books and shared in the business decisions. When needed, she would "punk whistle" in the woods, even while babysitting my brother and myself – which I remember well. During World War II she drove a GMC truck two trips per day the 60 mile round trip on gravel and dirt road from the mill to Coos Bay, hauling 4,000 board feet on each load. Robert E. Walls, in his Oregon Historical Quarterly article "Lady Loggers and Gyppo Wives" (OHQ 103, 2002) breaks new ground in describing the experiences of many gyppo wives in Washington, based on oral interviews. See also, parts of my Above the Falls. A student study by Courtney Berrien, "A Social History of Women in the Northwest Logging Industry from the Late Nineteenth Century through WWII," presented as a term thesis at Pacific Lutheran University, Tacoma, WA, (History 400, Presented to Martha Lance, December 16, 1998) is an example of the interest that a descendant of a "gyppo wife" may have in the role of her female ancestor who worked in the woods two generations back.

17. *The Timberman*, September 1953, p. 5. This is from one of 48 character sketches of significant logging personalities written by Stewart H. Holbrook and featured in advertisements for Bethlehem Pacific Coast Steel Corporation. They were all reprinted by Bethlehem Pacific in a booklet, "Personalities of the Woods: A collection of Bethlehem Pacific advertisements featuring personality sketches by Stewart Holbrook," no date.

18. For an introductory discussion of "Perfect Competition," the entry in the online free encyclopedia, *wikipedia.org* is a good place to start.

George Youst tie-mills in Cowlitz County, Washington
1927-1937

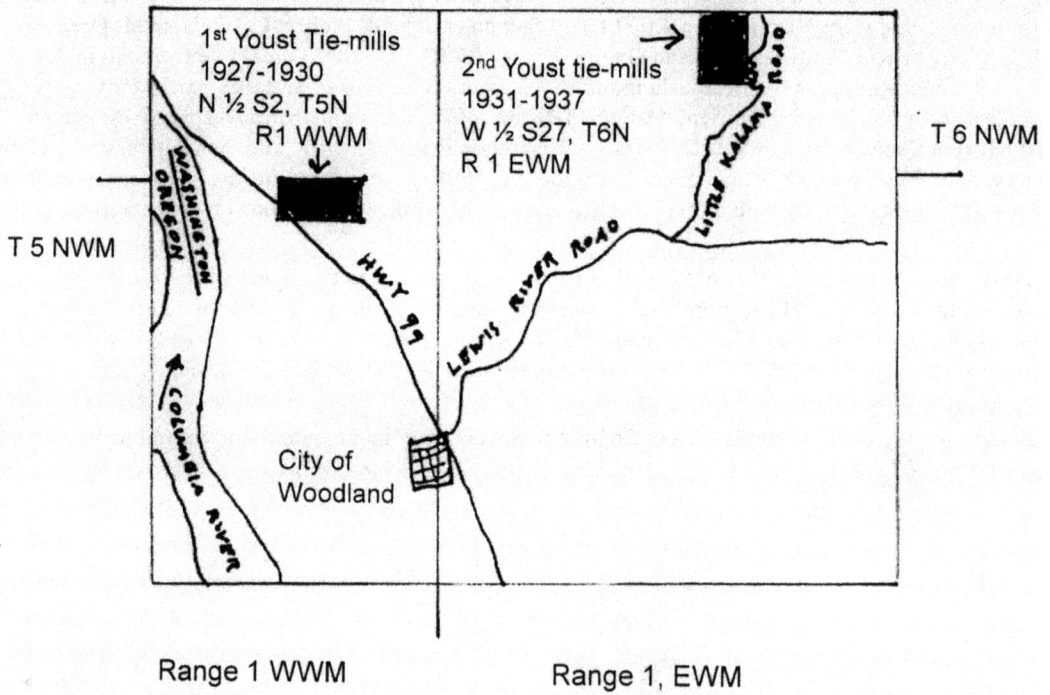

1st Youst Tie-mills
1927-1930
N ½ S2, T5N
R1 WWM

2nd Youst tie-mills
1931-1937
W ½ S27, T6N
R 1 EWM

T 6 NWM

T 5 NWM

WASHINGTON
OREGON

COLUMBIA RIVER

HWY 99

LEWIS RIVER ROAD

LITTLE KALAMA ROAD

City of
Woodland

Range 1 WWM

Range 1, EWM

The Tie Mill, Cowlitz County, Washington: 1926-1937

Introduction

My dad, George Youst, was born in 1899 and worked as a logger from the time he was 14. In 1926 he rose from the ranks of wage labor to the status of "gyppo" when he and a partner got a contract to load logs out of the Chehalis River onto railcars at his home town of Centralia, Washington. The equipment belonged to the company and the contract was to load the logs. How he did the work was up to him. This was "gyppo," in its original sense: contract labor in logging and sawmills in the Pacific Northwest. [1]

It was a start. The next step was setting up a small sawmill near Woodland, Washington, to manufacture railroad crossties, a tie mill. This led in 1930 to purchase of a tract of timber on the Little Kalama River, a tract of timber that helped him and his family to survive through the Great Depression of the 1930's. The transcripts from the interviews that appear below are presented as a document of a gyppo tie mill engaged in economic survival during hard times in the timber economy seventy and eighty years ago. It also serves as a case study of a distinct sub-culture that persists in the Pacific Northwest to the present day, the storied gyppo. [2] From the 1967 interviews:

Becoming a Gyppo: 1926-1930

Well, let's see. We got married in 1925 and the next year, 1926 I guess it was, me and another fellow [Earl "Curley" Rector] got a contract loading logs out of the Chehalis River. We had to bring the logs down the river, I imagine it was at least four miles around the curve in the river that we had to bring the logs down from the dump. Then we had to sort them and parbuckle the logs out of the water onto the bank.[3] The logs went to Olympia from Centralia. I think

Doris and George Youst, 1925

Doris Eagles Youst 1906-1982

George Youst 1899-1975

we got six-bits a thousand for loading. But we made pretty good money for that time. I know we loaded there two summers and I averaged $90 a week for the two summers.

"Six-bits a thousand:" That is, $.75/thousand board feet of logs loaded onto the railcars. If Dad and his partner each averaged $90/week, that computes to about 130 thousand board feet, or about 20 railcars of loaded logs, per week. Ninety dollars per week might not sound like much nowadays, but using an inflation calculator based on Consumer Price Indexes, it would take $1,056 in 2007 to equal the purchasing power of $90 in 1926. He thus averaged $4,225 per month in 2007 dollars, during the summer months for those two years, 1926 - 1927. His wages for loading logs when he worked for the Eastern Lumber Company at Centralia was $8 per day, which computes to about 2000 current dollars per month. Loading the logs during those two summers as a gyppo, he thus earned a little more than twice the pay he could have gotten for the same work at the going wage. The difference is that working for himself as a gyppo, he worked a lot harder, and a lot faster, than he would have done had he been working for wages at a large company.

H. H. Martin Lmbr. Co, 1924 one-log load. Mack McDonald, swamper, George Youst, head loader, Ross Riblet, 2nd loader

The owner of the timber, Doc Francis, was himself a precursor of the gyppo phenomenon and quite sympathetic to it. He had previously owned a sawmill at Galvin, a few miles northwest of Centralia, and he evidently owned timber in various locations. He was now logging his own timber using gyppo logging trucks, dumping the logs into the Chehalis River about four miles upstream of a main-line rail siding in Centralia. Dad explained, "It would be just a summer job, because that's when they could haul with the trucks. Then, I'd go out and get a job in the woods, where they had a railroad, for the winter." During the early 1920's, he usually worked in the winter for Eastern Lumber Company, previously the H. H. Martin Lumber Co., of Centralia.

I knew him [Doc Francis] before that, when I was a little kid I knew him when he had a mill out there at Galvin. Old Doc's timber is what it was. So old Doc, he got talking to me there and he says, 'fellows that work like you guys do, could make some money in a tie mill.' He talked me into investigating that. The next spring I got to thinking about this tie mill business. I didn't know nothing about a mill. But anyhow, I got hold of a guy who was kind of promoting the tie mills at that time, Bill Burke. He had a patch of timber down by Woodland and he took me down and showed me this. It was right on the highway between Woodland and Kalama. I could get this timber for a dollar and a half a thousand. At that time they was paying eighteen dollars and a quarter a thousand for ties.

3-log load, H. H. Martin Lmbr. Co. George Youst, head loader top rt. about 1924.

I talked to Curly [Rector] about it, about going in partners on the deal. I had a few dollars, not very many at that. Curly didn't have any. He had a car paid for, and I had an old Model T. I traded this old Model T in on a Chevrolet truck, a four-cylinder Chevrolet truck. Then, I got hold of a guy that had a tie mill up at Yelm [about 20 miles northeast of Centralia] that he wanted to sell. It run with a Fordson tractor. So I went up to see him about that. I forget what he wanted for it – I guess about twelve hundred dollars for the works. Just one saw was all that was with it, and the carriage and this old tractor. Feedworks was on the husk. I guess we had to pay him six hundred dollars down, and that didn't leave us any money. Curly had to sell his car. I remember that in order to get the car financed we had to take a horse in for a down payment on the car. So I think he got $300 or something like that. That wasn't enough to get us going so I got my Uncle Bill Bingham to go my

note down at the bank for $500. That's how we got financed to get started in this tie mill business.

To clarify some of the points my dad was making, a standard Douglas fir crosstie for the main line of the railroads of the United States and Canada was an 8-foot long timber, with a dimension of 7 by 9 inches. Forty-five to 50 million such crossties were used annually, most of them produced by small sawmills near the railroads, throughout North America. Other ties that were ordered from time to time included switch ties – used at the railroad switches. They were from 10 to 16 feet long and commanded a premium price. Ties for the sidings were smaller, usually 6 x 8 inches by 8 feet long. Foreign ties could be any of several dimensions and there were frequent orders for them, especially China. In fact, the Fairhurst Lumber Company specialized in orders for China ties, and claimed that over 2000 men in Cowlitz and adjoining counties were engaged in producing them even during the Depression.[4]

A board foot is a theoretical piece of wood twelve inches square and one inch thick. A standard 8-foot 7 by 9 crosstie therefore contains 42 board feet of wood (7 X .75 X 8 = 42). The actual unit of measure for sale of logs, lumber, timber, or ties, is the "thousand," 1000 board feet (designated as 1M, using the Roman numeral for thousand). 1M is about 25 standard 8-foot railroad ties, and delivered to the railroad dock at Kalama and loaded into box cars, a "thousand" was valued at $18.25 at that time. The usual order for railroad ties to the sawmill was by the "carload." A boxcar load of ties is 25,000 board feet, or about 600 ties to the carload, worth $456.25 – at that time. (To convert the 1926 dollars into current 2007 value, multiply by 11.74) Those are the units the tie mill operator was dealing in. Quality of the product was insured by sharp-eyed railroad inspectors who were quick to cull any ties that did not meet the railroad's very exacting standards.

The mill was a "No. 1 American," the smallest in the extensive line of mills built by American Sawmill Machinery Co., of Hackettstown, New Jersey, the premier manufacturer of small sawmills at that time.

The parts of a circular sawmill need some explanation. When Dad says, "Feedworks was on the husk," there are probably not many readers who would know what he meant. Without getting too technical, the husk is the frame on which the

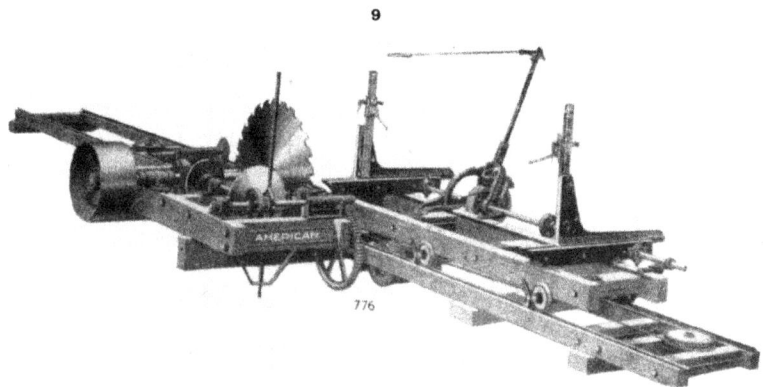

American No. 1 Saw Mill
With Variable Friction Feed
Right Hand Mill—Code Word, Wocdu
No. 1 Mill

mandrel of the saw is mounted. The "feedworks" is the mechanism that "feeds" the log on the carriage into the saw and returns it for the next run through the saw. This mill used a variable friction feed that permitted variable speeds of the carriage, in either direction. By the mid-1920's the feedworks of most small sawmills was a more reliable variable belt feed mechanism. From Dad's description that the mill consisted merely of the Fordson tractor engine, one saw [and its mandrel], along with the carriage and its feedworks, we have the bare essential parts needed to manufacture lumber, and nothing else. No frills. But even those bare essentials posed an insurmountable mystery to my dad at that time. He explained,
:

> So, we took this little old mill up there. Then I had to get a guy that knew
> how to saw and set up the mill. This guy that got the timber for us [Bill
> Burke], he knew about this sawyer up at Yelm. I called him up and talked to
> him and told him that I had to have somebody that could do the whole works
> about the mill because I didn't know nothing about the mill. I can get the
> logs there. So I says, "How much would you have to have for that?" "Well,"
> he says, "while we're setting up the mill and moving, it will be six dollars a
> day and while we're sawing it will be eight dollars." That was satisfactory.

You know, we went up to Yelm and we loaded that mill up on Monday morning, and Saturday we sawed our first tie. We sawed our first ties on Saturday! Moved it down, set the mill up, and the whole caboodle and when we went by the broker that was gonna buy the ties, he lived at Napavine, he says, "When you gonna get started?" "We're already started! We sawed ties today!" He couldn't believe it.

We had to deliver those ties to the Port dock in Kalama. We sawed lumber and built three shacks. I had a shack and Curly had a shack there. I hired a guy to log with a team. I paid him eight dollars a day for him and his team. We got by on the first setting, then this guy with the team got so independent that he came over after his money – the ties were shipped but it would be about three days or so after the ties was shipped before we'd get our money. He come over and boy, he was demanding his money! I had my pencil out and figured how much he had coming – how many days and he got pretty sassy. I put my pencil in my pocket and I reached in to drag him out and he got his feet up and "no, no, no!" I was gonna drag him out of there and work him over. I told him, "I've taken more guff off of you than anybody I ever worked for, trying to humor you, because you had those horses. Take your horses home. We'll have horses of our own here tomorrow. As soon as we get our money for these ties we'll have our own horses."

Earl Rector and Geo Youst, horse logging to the tie mill near Kalama, WA, 1929

Setting: The small tiemills were set up in the timber. The timber was logged directly to the mill, frequently with a steam donkey, but in my dad's case when he first got into the business, by horse logging. When all the timber surrounding the mill for 1500 feet or so was logged off, the mill had to be moved to a new "setting," with a fresh supply of timber. One setting of a small

18

semi-portable tiemill usually equaled about 40 acres of timber, and the "forty" is the usual unit of purchase for timberland. Four forties equal a quarter section, which might mean four settings of the mill. A full section is one mile square, and contains sixteen forties. An interesting article in the September, 1930, Timberman described the operation of the typical tie mill: Timber was felled and taken directly to the mill. They employed 4 or 5 men, sometimes using family members. In one mill at Morton, the women of the family worked in the mill with the men. Some of the mills used steam logging, some used horses. Some logged one day and sawed ties the next, using the same crew for both. Typically they were in 60-year-old second growth timber. In the fall of 1929, there were 60 tie mills cutting ties for Fairhurst Lumber Company in the area around Morton, Washington. The next year, 1930, there were only 20 mills remaining due to the emerging economic depression.[5]

When we got our money we went down to Portland to the Cow and Horse Exchange down there. I didn't know nothing about horses but Curly knew about horses. He picked out a team. He got one good horse and the other one wasn't much good but our money, we could only go as strong as our money would go. So we got one good horse and one poor one, but she was willing to work. She'd do all she could. I think it only cost us $160 for the team and the harness. If we'd got good horses, it would probably cost us about $250 for a good team. The old harness was chain tugs. They had built that dike and things around Longview with horses. When they got through with that, there was no need for the horses and the harness, and [the Portland Cow and Horse Exchange] bought a lot of that stuff from there. Anyhow, we took 'em and we didn't have to hire that guy again.
He was a nephew of the guy that owned the timber that we was logging.

When we got our own horses, that was a little better. Curly drove the horses and I worked at the mill and done everything there was to do and then hauled the ties. So after awhile Curly, I guess the third setting it was, Curly

19

he decided I had the easiest job. He wanted me to drive the horses. He'd haul the stuff [drive the truck]. And when he hauled, that's all he did do. So the partnership wasn't going as good as it could have.

That's the place where we had the old horse, Dick. He'd sooner do anything than work, but he liked to eat. One time we was working with a singletree – just a tong and we'd hook each horse onto an eight foot log, and haul it in. One day old Dick, when we turned him around to go back to the woods, he took off for the barn. So I took off after him. He had this tong dragging behind him, and I was hollering at the old son of a bee, to stop. But he didn't stop, he just kept on going. He was speeding up a little bit, and when he'd jump over the limbs the tong would fly and hit him on the rear and would make him go faster. He went on over to the barn, and when I got over there he was just standing there, fine and dandy. So I got on him and rode him right back and put him to work again. We had a lot of trouble that way.

We had a setting there that had an ungodly pile of slabs piled out. Beautiful slabs. We had enough timber left to make two more carloads of ties, and it was getting a problem getting rid of the slabs. So Curly decides that we burn them, get room to pile the rest of the slabs. I, like a nut, went along with him and we set them slabs afire. The mill was setting in amongst great big alders, they were big, straight, tall. The mill was setting right in amongst them alders and he set this thing afire and the heat from those slabs was so intense those big alders, you'd see them start buckling like that, and break and fall from the heat. Of course we just shot the sawdust right out alongside of this old tractor. The fire would get into that sawdust, and pretty soon it would be into the mill. But we had a water hole there, and by carrying buckets of water and dousing on that, run in there and douse the water on it. All night we was fighting it. It charred the skids under the end of the mill, but we

saved the mill. And we got rid of the slabs!

Slabs: Slabs are the rough, outside pieces sawed from the log and usually burned as waste. The slabs generated from a mill exclusively cutting railroad ties contained an enormous amount of perfectly good wood which could have been cut into dimensioned 2-inch lumber had there

Back end of a typical tie mill of the period, 1920's and 30's. Note sawdust conveyor and sawdust pile at right. At left, stack of ties and 2-inch side cut. From Timberman, May, 1934, p. 16.

been a market for it – and if the mill had an edger to saw cants into dimensioned boards. In the case described above, the "beautiful slabs" were the full side-cut removed in "squaring" the log for manufacture into ties. Careless disposal of slabs by burning was the cause of many small sawmills going up in flames, over the years.

When we got through with that setting, that was the end of Curly and I. I made him a price. I'd sell my interest to him for so much or I'd give him that much for his interest. He was gonna buy me right away so he went up to Chehalis and went to all his friends to borrow the money to pay me off, but after a couple weeks he couldn't make'er so I bought him out. Probably fifteen hundred dollars, something like that. That meant the horses, the truck, and the whole caboodle. Then, he wanted the sawing job. He wanted a contract for sawing and hauling. I let him go to work on that. I had fallers, and I paid them so much a tie. I think seven cents a tie, I give them for falling and bucking. Seven cents a tie I believe is what it was. Eight feet, then logged'em with the horses. I was doing the logging. When they had the first tie inspection – they'd only come in after you got a whole mess of ties down there, they'd bring the inspectors and the turners to turn the ties. I

think that about 20% or something like that of the ties Curly had sawed was culled. Mismanufactured. He was getting so much a tie for sawing and hauling. He wanted his money right away so he could get out of there. He says, "When them fallers find out about all them culls they'll kill me!" He wanted his money to get out of there right away. I paid him off and got him out of there.

The bill of sale sates that Earl Rector of Cowlitz County Washington, "grant, bargain, sell and convey" to George Youst "all my one-half partnership interest in and to: one No. 1 American Tie Mill, one Fordson used as power for same, Three horses, all logging tools of every nature and description, and all timber in section 2, Twp. 5 N. Of R 1 W of WM., in Cowlitz County, Washington, now owned by Youst and Rector, a co-partnership composed of Geo. Youst and Earl Rector." It was dated February 19, 1930.

Falling and bucking: In the Northwest, "felling" a tree is spelled and pronounced "falling." The men who do that work are "fallers." In the rest of the English speaking world they are called "fellers." Bucking means cutting the fallen tree into logs of a specific length, in this case, eight-foot logs. The men who do this work are called "buckers." (Small logs are "bucked" into firewood by placing them in an "X" shaped sawhorse, called a sawbuck, hence the term "bucking.") Falling and bucking are the two tasks of the "cutting crew" of a logging operation, as distinct from the "rigging crew," which is responsible to move the logs to a landing by whatever power source is being used. In this case, it was horse power.

That old tractor was getting awful, awful poor. You couldn't hardly get it started in the morning. We'd get everybody on the belt and one guy on the crank, cranking it and turning. Boil the oil to get it hot, and all that kind of stuff. Sometimes it would be 10 o'clock before we'd get the darned thing started. I decided I couldn't put up with that so after I got Curly out of there I went up and bought an old car at Steelman's at Chehalis, a six cylinder – stroke was 4 by 6 – I forget the name of the car. Anyhow, I put that in there

and that was a lot better. I got old Dude, Dude Ogle to come over. He was sawing for another outfit there, and they couldn't get logs enough to keep them going. So Dude came over and sawed for me.

This was Dad's first major upgrade in the power source for the mill, replacing the old Fordson tractor engine with the more powerful, and more reliable, 6-cylinder car engine. It was internal combustion engines – first gasoline and later diesel – that made semi-portable sawmills possible, and they revolutionized the manufacture of railroad crossties. Prior to the proliferation of these small sawmills, most railroad crossties in North America were hewn by hand from any of about 26 different species of tree acceptable to the railroads. The railroads preferred crossties that were sawn for several obvious reasons, primarily having to do with cost, standardization, and grading. In the Douglas fir regions, virtually all crossties were sawn.

Water powered sawmill, Tillamook, OR, 1930's. Family operated, not portable!

The earliest sawmills were anything but portable, being water powered and totally site dependent. By the turn of the 20th century water power had been almost completely replaced by steam, and there were attempts to make steam sawmills that were comparatively mobile. As late as 1928, Western Farquhar Machinery Company in Portland was advertising "sawmill rigs mounted on wheels or sills. Quick steam. 25 to 50 horsepower." But a steam engine needed a year-round water supply that could be piped to it. Small sawmills set up in the timber may or may not be near such a water supply, which limited the practical use of steam for that application. The one perceived advantage of steam – utilizing waste wood from the mill for its fuel – had the disadvantage of needing extra labor to fire the boiler. The very low

cost of gasoline or diesel, and the fact that the internal combustion engine had no site-specific limitations, soon rendered steam and water power obsolete for the small sawmill.[6]

A "Timber Baron," 1930-1937

When we got through with that, then we bought that half-section of timber up in the Little Kalama.[7] That was advertised in the Oregonian. There was an outfit in Vancouver that made coat hangers, Acme Coat Hanger Co. had it advertised. The bank in Woodland was carrying the paper on it [Security State Bank, C. A. Button, President]. They had bought it for the alder that was on it because that's what they used to make coat hangers out of. Alder. So I went up and looked at it and then I went down and saw the guy that owned it, Acme [L. L. Dillon, President]. The panic was coming on then, that was in 1930. Things was getting pretty tough. He was anxious to get rid of it, because he had to make payments on it. He made me a real good deal, I think it was only $2500 for a half-section. Yeh! Twenty-five hundred dollars! I had to borrow some money again at the bank, to make a down payment on it. Then I had to pay a dollar a thousand as I took the timber out.

So, the price of ties kept going down. From eighteen and a quarter [$18.25] they went down – the last two carloads that we'd cut on the Burke's place, we got twelve and a half [$12.50] for them, at the OWR&N.[8] By the time we got up there [to Little Kalama] there was no market for them. Once in a while Fairhurst would get a little bit of order for some ties and he'd divide it up amongst all the mills. There was a hundred and twenty-five different mills cutting for Fairhurst. He was the big broker in Tacoma. Two different orders he got there, my part of it was 25,000 [bd. Ft], just a carload, on each of two orders, and it was a long ways between orders.

Fairhust Lumber Company of Tacoma, Washington was operated by the brothers A. W. and C. J. Fairhurst, and C. J.'s son, Jack. It was a wholesale firm, actually a brokerage firm, specializing in "Export and Domestic Railway Ties and Lumber." The function of a brokerage firm such as Fairhurst was to secure lumber and ties from small mills to supply large orders, foreign and domestic, from railroads, retail yards, and governments. Without the brokerage firm, the market for the output of small mills would have been limited

CHINA TIES: 6,500,000 board feet of ties being loaded for China aboard the S.S. San Julian in Portland, Oregon. Photo was the cover of April, 1934 issue of Timberman.

to local demand. Large orders for the collective output of the small mills would have been very hard, if not impossible to fill. The Fairhurst Lumber Co. also invested in sawmills and timberlands in western Oregon and Northern California.

Horse Logging

We had quite a lot of trouble with the horses. I hired an old farmer to help log. He brought his team up there but he didn't know the first thing about logging. He had an old set of harness that was pretty near as old as he was, leather harness. He was constantly fixing tugs, this that and the other. More time fixing than he did hauling. Finally he pulled a sixteen foot log up just barely by a tree then he turned, right against that tree, went through the harness again! I said, "Jim, you just as well take your horses home because you can't do me a bit of good here." But he was a good old guy. He was one of the old timers up there.

Horse logging in Cowlitz County, 1929. Geo Youst, left, his teamster, right (with peavy) – manhandling a log through the dirt onto the rollway of the mill. One horse behind him, one horse at extreme right of photo.

One day, I was logging and it was hotter than heck. We was pulling logs up out of the creek. Them horses, I'd just turn 'em loose when I'd feed 'em at noon. They'd just feed and when I'd come to take 'em out in the afternoon they'd be there. But one day old Maude, she decided she didn't want to work that afternoon. She went on strike! She took off up the road. I eventually got in front of her and drove her back down to the barn and when I got back down there it was no trouble to catch her. I didn't put old Dick along with her. I made her work all by herself that afternoon. She put in the afternoon doing what both of 'em would have been doing if she hadn't done that trick. I evened up with her a little bit that time!

One night I heard them kicking out in the barn. We built a nice barn for them. I heard them kicking out there and I finally went out. When I got out there, Maude had kicked old Dick in the leg and broke his leg in two. He was down and the bone was out through the meat and skin, laying there. I had threatened to kill the old son of a bee many times before, and cussed him with everything I could think of for different things. But when I saw him laying there I felt sorry for the old bugger and I didn't have the nerve to knock him in the head. I went over and got Dude out of bed. Dude had a 25.20. I got him out of bed and got him over there. I stayed outside the barn while Dude shot him. I'd taken Maude out and put the harness on her while Dude was getting the gun and shooting him. I got the harness on old Maude and I throwed a choker on Dick's hind legs and drug him out of the barn. Dude says, "Be easier to drag him across the road, down in the bottom, in the woods down there." That's what we did, and that's one of the biggest mistakes we ever made. When that thing started to stinking, Holy ol' Kokomo, then I wished I didn't smell. The flies, them big bull flies, it got to stinking so bad we couldn't get close enough to him to put lime on him. Oh, it was so terrible. But we had to put up with it till all them maggots had eat him up. That was really something.

We got rid of all the horses. We didn't have any logging equipment then. Logging equipment was all gone. I had horse collars, though. A fellow halfway down the hill that had a place there, that summer he put hay in the barn. One load of hay and the barn caught fire. But he had insurance on this old barn so he got the insurance, but it burned up his harness and stuff that was in the barn. He couldn't very well get that out and get his insurance too, with just one load of hay! He heard that I had harness and he wanted a horse collar. I had one new collar and he wanted to know what I'd take for that collar. I says, "I paid six and a half for the collar, I'll sell it for three and a half." So he gave me three and a half and that was the first cash money we'd had in a long time. What did we do but we jumped in the Ford and took off for Centralia and went up to visit.

It is surprising how much of the logging in the Pacific Northwest was still being done with horses even as late as the 1930's. It was the collar harness, in use for a thousand years, that made horse logging possible, but until about the turn of the 20th century yoked oxen were the motive power of choice for most loggers. Horses were faster and could work longer hours, but they ate oats, were more expensive, and they needed good harness. The master maker of logging harness and horse collars was John P. Sharkey (1870-1939). He came to Portland in 1885 and from his shop at 605 NE 21st Ave he advertised that his collars were "best for lumbering and logging, guaranteed to give satisfaction." They were hand made, and stuffed with "long straw." Sharkey horse collars, said to be the strongest horse collars made for logging and lumbering, were known throughout the Pacific Coast wherever horse logging was conducted. He also made muleskin lumber-piler and teamsters' aprons of the "highest quality." He died in 1939, by which time horse logging was pretty much a thing of the past.[9]

The Depression

The panic started getting worse and worse. Couldn't get no orders and I had this Ford truck paid for and had just bought a set of chains, because there was no gravel on the road at all. I bought this new set of chains for the hind tires. I took it in and had some work done on it, I think it was fifty-six or fifty-seven dollars worth of work done on the truck. No orders, and so I couldn't pay it. So I told the guy down at the Ford garage, "By Gollys, I just don't have the money to pay for this." He said, "can you pay anything?" I said, "I couldn't guarantee that I could pay fifty cents a month. I couldn't

guarantee that." So I says, "I'll tell you what I'll do. I'll bring the truck down and park it right here and you sell it to somebody else." And that's what I did. And I left those brand new chains and they was twelve and a half ($12.50), and they was paid for. I left the chains hanging right on 'er.

It got down to where we couldn't even get money to buy gas with, and you could get gas for ten cents a gallon, by the barrel. We had a setting we were pretty near through with. We would cut ties and pile them along the road – we had about twelve hundred feet of plank road up on our place. We had ties piled on both sides of the road, but no orders for them. We wanted to get that setting out so we could move the mill. Dude figured that with a hundred gallons of gas we could cut out the rest of the setting. So he went down to the bank and tried to borrow $25, so we would have enough to finish this setting. And the banker wouldn't loan him $25! He come back and he was so mad. The next day I went down and I told the old banker my story and he loaned me a hundred dollars. That made Dude still madder, to think he would loan me a hundred dollars and wouldn't even loan him twenty-five. So we had plenty of money to finish the setting.

Geo Youst, lft; Dude Ogle, rt. 1933, hard times in the tie mill business.

Finally, we got an order for mine timbers. We was cutting standard size ties, is what we was cutting [7 by 9 inch]. Then when the piece was too small, we'd make a 6 by 6 out of it. We'd cut them too, and pile them out.

So finally, the snow was on and everything was froze up, and we got an order for 6 by 6's for mine timbers. We loaded them 6 by 6's up and hauled 'em down there and got the cash for that and Dude and I split the money and we jumped in the car the next morning and right down to Portland to Montgomery Wards and we bought a great big radio. A ten tube radio, took three batteries to run that thing! We had music then!

10-tube radio: If I remember correctly from overhearing my parents talk about it, the radio was a battery operated 10-tube Atwater Kent. The mill was only about eight miles from Woodland, but it was a dirt road which went over the divide from Lewis River to the headwaters of the Little Kalama River. It was quite a remote location at the time, and there was never electricity during the seven years we lived there. Atwater Kent made several models of battery operated radios, but the 10-tube model took so much power that the batteries were frequently dead. I don't believe my parents got as much good from the radio as they had hoped.

The county had lots of bridges that was rotted out and so I got quite a few jobs cutting planks to replace the county bridges. I'd get pretty good money for that. I'd get about twelve dollars for that. We cut ties for as little as six dollars and eighty-two and a half cents a thousand to us, delivered on the dock at Kalama. But the plank I cut for the county, I'd get twelve and a half ($12.50) a thousand for 'em, but I'd have to cut all the different dimensions that they wanted. And I only had an eight foot carriage.

One bridge down by the school house, they wanted thirty-six foot long, 12 by 15 stringers to go across there. There was five stringers to go across, that long. Dude and I cut them on that little eight foot mill. We picked out logs that would just make it. I had bought an extra mill, and I had the [extra] eight foot carriage with the wheels on it, and we fastened that on behind the [original] carriage so it would balance. We fixed the rollway so it was long

enough so we could roll them long logs down there, and we'd set'em as far ahead up to the saw as we could when we rolled 'em on, and the carriage way back there, and then Dude would run it through as far as he could, then back up and we'd chop that piece of slab off. Then we'd have to turn it down and put a block on the bunk, underneath where the slab come out of, to hold it up. We'd go clean around, four slabs off the front end [of the log], and then we'd roll it off on the deck and back the carriage up to catch the back end of it.

Eight-foot mill: The small sawmills were commonly designated by the length of the logs that the carriage was designed to handle. Standard railroad crossties were eight feet long, and a mill cutting primarily crossties, a tie mill, would only need to have an eight foot carriage. That presented special difficulties when orders for longer pieces would come in. To cut 36 foot bridge stringers in an eight foot mill took a lot time and ingenuity, as described above.

Steam Donkey: 1931

We was still out of any logging equipment. I saw in the paper that there was an outfit in Portland that had a donkey for sale up Lewis River. We went up and looked at it and it was a 10 x 12 Seattle three-drum, pretty good rig. The frame was broke. Cast iron frame, an old donkey. I went down to Portland to this big office where this outfit was and started dickering for this donkey. I forget what they wanted, they wanted something like what the donkey and the rigging would be worth. I wasn't in any position to pay that kind of money for

George Youst's 10 x 12 Seattle steam donkey with the new sled he built for it.

31

it. So I kept cutting them down and finally I got it down to where I figured I could raise the money to pay it, I guess $750 or something like that. They started out about $1500. I got 'em down to about half. $750. I think at that time I had $150 of my own. Then I started talking terms on the rest of it. We come out that I was to give them a dollar a thousand as I cut, for the rest of it.

10 by 12 Seattle: Steam donkeys were hoisting engines designed specifically for logging. They were designated by the size of their piston and where they were built. A 10 X 12 had a piston with a 10-inch bore and a 12-inch stroke. I have the bill of sale for that donkey, purchased from Union Steel and Rail Company of Portland on May 15, 1931. The Seattle was built by Washington Iron Works in Seattle, and this one could have been built any time between about 1902 and 1913, when they began using a cast steel frame instead of the cast iron frame. It was a "three-drum" donkey. The three drums were the main drum, for the mainline which pulls the logs in; the haulback drum, for the smaller line that hauls the mainline and the rigging back into the woods for another turn; and the third drum is an auxiliary drum for the "strawline," a smaller line easily pulled by hand and used to string the larger lines into position for logging.

I'd been figuring just how I was going to work this deal, so when I come out of there I started talking to Dude into buying the mill. Dude had forty acres of land out there by Klaber. I started talking to Dude to sell him the mill, then I could take the cash – I figured he could borrow the money on that property and I could take the cash and pay [for the donkey and rigging], and have a little bit left to do what I wanted to with it. Dude had never thought that the property would be any good for collateral. So, I talked him into going up there and in a few days he came back with the cash.

I had to go up and take the rigging out of the tree. And that was a raised tree, and I was foolish enough, I took the doggone guy lines off that SOB, and all

32

that stuff, and it never fell. It had been setting there long enough, and had built up debris around it. I thought it was a tree, but it was a raised one! It just stood there, and I went up and let the guy lines down. Had to let the blocks down and that kind of stuff. It could have fell.

Raised tree: High lead logging, using a spar tree to keep the rigging in the air and give lift to the log coming in, was in general use in the Douglas fir region of the Pacific Northwest by about 1915. In the older ground lead, the lines were strung from the donkey engine along the ground, resulting in excessive friction as the log coming in dug into the ground, and frequently hung up on stumps, rocks, and other logs. The answer was to hang a bull block high in a carefully selected tree supported with guylines as much as 200 feet in the air. If there was not a suitable natural tree standing where the spar tree was needed, a tree-length log could be raised perpendicular, secured with guy lines, and used as well as a natural tree. When the setting was complete and the rigging was to be removed from a raised tree, the guylines were taken loose from their stumps on the ground and the raised tree would fall. In the case described above, the tree – which had evidently been there a long time and had debris piled in around it – did not fall when the guylines were cut loose, and my dad assumed it was a natural tree. He went up the tree to remove the blocks and guylines from it, as he would have done had it been a natural tree but in this case nothing was holding the tree up. He was very lucky that it did not fall and kill him.

We had to move the donkey out of there. We had to take the boiler down to haul it out, then had to put it back together again. It was at a mill, and there was only a little bit of pipe. We got the rigging, but we didn't have water pipe enough to gravity water down to it. This same outfit had a logging camp up in the woods. They had all kinds of water pipe up there. We went up one evening, we was gonna take up this pipe. We didn't have any right to take that much, but I did have it on my sales slip that I was entitled to the pipe that was there at the donkey. We went up there and we kind of felt like we was

stealing it, which we was. There was a couple fellows that run cattle up there, and there was all kinds of guys killing beef, during that time. They talked to us and we went down to see where the pump was. We came back from there and started taking up that pipe and we had the truck parked the closest place we could get across from this railroad track to pack the pipe across there. Them fellows were parked by the truck. We gathered this pipe up and had it all taken up, where we could take it across, but those guys just stayed there. Finally, I knew we was caught so I took a couple pipes with me and walked across. I told 'em I'd bought that donkey down below and there wasn't enough pipe to run the water to it and we was up there picking up some pipe. "Oh," they said, "We thought maybe you was after a beef." So that gave us enough pipe to get water to the donkey.

Besides having to take that extra pipe from that outfit, we didn't have a water tank. There was an outfit had a big camp out of Kelso way back in the hills. There had been a fire come through there I guess in 1930 or 31. It burned out several of their trestles. Then the Panic hit, and they just left the machinery set there. They didn't have the money to build them trestles over again. They had a cookhouse with a nice round water tank for the cookhouse. So, Dude and I went down there and took the water tank apart, took it all to pieces, and packed it up the hill – oh, it was a long ways to pack it before we could get it to the car. I imagine probably two miles. We'd pile up as many boards as we could pack, and pack it up that hill, and loaded it in the back of Dude's seven passenger Studebaker, hauled it home and put it back together, and we had a tank on the back of the donkey.

Bill [Yost] came up and helped us build a sled for it.[10] We got that going, then it wasn't much trouble getting logs. It was a pretty good rig. Steam. We yarded as high as 1500 feet with it, you might as well say on the ground. All we had was second growth there, and we could only get up maybe about a hundred feet, about all the lift we could get. Get the lines off the ground. So I had Bill work there for awhile, and Swede [Yost] was there awhile, and Frenchy was there. He was Bill's age. He worked in the woods, he was working up at Union Mills when I was loading there. We got it lined up and got the tree rigged and started logging. We'd just get orders now and then, that□s how it was all during that Panic. Dude and I finally got down to where just two of us worked. We'd fall the trees, then yard them in, and then if we didn't have an order sometimes we'd go ahead and cut up standard size ties and pile 'em out.

Longshoreman Strike: 1934

Then, we got an order for some mainline ties, delivered to Longview, about the time the Longshoremen had their strike. That was the first big strike under [Harry] Bridges. Of course, we was loading on the main line. We didn't have nothing to do with the [Port] dock. But you know, those guys came over and picketed the [tie] docks, and wouldn't let the trucks on them. On top of that they would sit there all night and burn ties. Take them ties and burn 'em. That's how they kept warm, take them ties right off the pile and burn 'em. So, you know, Bridges didn't make much of a hit with me on that deal. Finally, they got the strike over and we got to shipping the ties again.

All of the ports on the West Coast were involved in the strike of 1934, including the Port of Longview.[11] My dad's reaction to the strike, and to Harry Bridges as its head, is instructive. The Longshoremen were organized as a radical union with the slogan, "An injury to one is an injury to all." The solidarity of organized workers was paramount, and the gyppos, as independent contractors, were not organized. The essential dichotomy between the unions and the gyppo went back to the lumber strike of 1919, when the "gyppo system"

35

was introduced to help break the power of the union in logging camps and sawmills.

The International Longshoreman's Association (ILA) strike of 1934 was the first industry-wide strike in maritime history, ultimately involving 35,000 workers from San Diego, California, to Juneau, Alaska. It lasted 83 days during the summer of 1934, and was characterized by violence in the major ports of San Francisco, Portland, and Seattle, with several deaths and many injuries. Despite rabidly anti-union reporting in many of the national newspapers – especially the Hearst newspapers – the general public appeared quite sympathetic to the strikers despite isolated unlawful acts such as burning the ties to keep warm on the picket line.

The strike resulted in significant gains for organized labor, which had been legitimized under the National Recovery Act (NRA) of 1933. Harry Bridges (1901-1990) was president of the ILA in San Francisco at the time of the strike and emerged as its leader, achieving national notoriety as a result. In 1937 he was instrumental in forming the successor International Longshoreman's and Warehouseman's Union (ILWU), which has proven to be an extremely effective union. He headed it for forty years, until he retired in 1977.

WPA: 1935

Louie Skinner brought his truck up to haul lumber for me.[12] He'd bought forty acres of land the other side of where I was at. He lived there, and he had four kids, and no work. Then Roosevelt got elected. They put on that PWA. They put Louie in charge of the crew that was working on the road. Once in awhile Dude and I would get a little order for some ties and we'd cut 'em but the ruts got so deep we could only get up that hill with about 500 feet of ties. There was no gravel on the road whatever. Louie was working his crew and finally, about the time Dude and I got an order for some ties, Louie decided that them ruts should be filled up. He comes up with the crew and shoveled that muck off the side of the road into the ruts, and it was slick as heck when

36

he'd fill that mud in there. Of course that mud wouldn't stay. The first truck through, if you could <u>get</u> through, would push it out. So finally, I stopped and told Louie, "Take them guys somewhere else. We're the only guys in here that's making it on our own, and you come up here and try to sprag our wheels by throwing mud in the ruts so we can't get out. So he took his guys.

WPA / PWA: There is a common confusion between these two major New Deal programs, both of which were designed to lessen the effects of the Depression. PWA, The Public Works Administration: "Big bucks for big projects," provided matching Federal grants for large construction projects – e.g., the Oregon Coast Bridges. The WPA, Works Progress Administration: Louie Skinner was employed by the WPA, created by Executive Order in April, 1935 and funded by Congress in July. It's purpose was to provide employment and income to one wage earner in each unemployed family for a maximum of 30 hours per week at the prevailing minimum wage in the area. By 1938, 3.3 million of the 20 million people on relief were working in WPA projects, making it the largest employment base in the United States. The standard Conservative criticism of it was that it wasted money on projects that were neither wanted nor needed - such as filling mud into the ruts of the County Road at the Little Kalama. My dad never tired of using that story as a prime example of the folly of New Deal programs. By the time he was of retirement age, however, he had lost his sawmill in a general recession in the timber industry and was grateful enough to receive the benefits of one New Deal program, Social Security. The older he got, the better the New Deal looked to him!

Portable Sawmill: 1936

We cold decked a big bunch of logs in the canyon, then moved the donkey back to the mill. Just got it moved to the mill and a fellow came along and wanted me to take the mill up to Stevenson, on the Columbia River. He

37

had some timber up there he wanted cut. I had sold the mill to Dude by then. Dude didn't want to leave until we got the cold deck sawed up and I said go ahead [take the mill to Stevenson]. He said, "What are you gonna do?" I said, "I'll get by." So, there was an Advent at La Center who had three or four mills he had built on wheels, with a Studebaker 6 to run 'em with, and for the axle, he had the arbor made on it. A dandy set up. I went over and talked to him and he was just about through on a setting and so he said, Yeh. He'd come over and saw 'em out for so much. He finished up over there at ten o'clock one morning, a few mornings after that, and by two o'clock that afternoon he was ready to saw over at our place. That was really portable. You know, he did pretty darned good too. I run donkey and hired two of his men. One to hook the logs on out in the woods and the other to run the drag saw at the mill. Them fellows stayed and cut out that cold deck. They averaged four dollars a day there, all by the thousand. They were tickled to death. That was a lot of money then. The best money they'd made.

The role of the Seventh Day Adventists in developing small sawmills and portable sawmills is a subject worthy of further research. Very often a successful small sawmill operation – and sometimes a large one – would be known as an "Adventist Mill," owned and operated by Seventh Day Adventists. This was true throughout the sawmill regions of Washington, Oregon, and California and perhaps elsewhere. Whatever the social, cultural, or economic forces behind the phenomenon, it was an interesting fact that, to my knowledge, has not been explained. Portable sawmills, those on wheels and capable of moving from one setting to another in a matter of hours, were fairly common in the southern and eastern states but were practically unknown in the Pacific Northwest. The Seventh Day Adventist mills described above were the first truly portable mills my dad had ever seen.

That was all the timber I could reach with that one donkey, to that setting. I still had about 200 acres of timber left on that half section, but I bought another 160 acres [of timber, not land] up on the hill. I moved the donkey up there and I got some of his men [the Adventist] to come to work for me. We moved the donkey up there and rigged up and I went up to Mt. Rainier to buy another mill. Bill Yost went up with me. Oh, it was cold, freezing that day. I had a four cylinder Dodge, a "Fast Four." I think it was a 1928 or something. We went up there and bought the mill and hired a truck to go after it. It was on a sled.

During the 1930's used sawmills and sawmill equipment was easily available from operations that had failed due to the economic depression. One ready source of sawmill and logging equipment that my dad availed himself of repeatedly over the years was the Alaska Junk Company (Schnitzer Machinery Company) at 1st and Taylor in Portland. Sam Schnitzer (1880-1952) started in the junk business in Portland in 1906 and by the 1930's he was buying the second-hand machinery of entire sawmills, logging camps, and railroads. He advertised, "Sawmill machinery and equipment. We always undersell."

My dad established a good business relationship with Mr. Schnitzer and found the Alaska Junk Company a necessary stop almost every time he went to Portland all the years he operated sawmills.

We got the mill down and went to work with it. We cut out about 80 acres off that four forties that I contracted for. I got Hank Wesfall and his brother. Hank was a guy I ran around with, went to school with. He'd got so bad, done such awful things before, that he finally got religion down there at Portland in one of them holy roller outfits. He got religion. He come up there and went to hooking for me. We was rigging a tree and I was running the donkey. I had the guyline tight and was holding it, and pretty soon I could hear the hammer hit them spikes out there and pretty soon I heard the awfulest cussing you ever heard. When they got the guyline spiked up and Hank came in by the donkey, the blood was running down the side of his face. One of those spikes had hit him right by the eye. I says, "Can you see the light, Hank?" He says, "I can see 'er, but I can't walk in 'er."

Hank and his brother was living in a little shack we had built just above the mill. Hank would go down and fire up the donkey and have it hot when I got down there in the morning. One morning I went right on by the shack and down to the mill, and no fire in the donkey and so I fired it up and filed the saw, and nobody showed up. I went back and I saw outside the shack where they had been throwing up. Hank says, "We went down to Rosy's last night. Old Rosy got out the Vino, we had a few shots of Vino. Rosy got out his old accordion that was all patched up with adhesive tape, to cover up the holes in it. He wousled it around and he got it down between his knees and he squeezed out "The Old Spinning Wheel." We had quite a celebration. So even a preacher can fall from grace!

We logged that one side of the tree, and the power company was coming through with the power line. They wanted to come through right over where we was logging. The guy that owned the timber wrote me and asked me if it would interfere. I wrote back and told him I didn't care if they brought the line over if it didn't interfere with my logging. I had the hillside all fell, and we was logging, and they come down through there with a whole crew, about fourteen men pulling this half inch copper line down through there, right over our logs, right where we was logging. Right over the top of our line. Well, we couldn't, while them guys was in there, we couldn't log. So we had to lay off for a few days until they got that strung and the wire tightened up.

Hank, he was hooking and I was running the donkey. He was out there on the hill and our line wouldn't raise as high as the power line. It was about 1800 feet between poles, across that big canyon. But a log run wild and went off sideways and pulled a small tree that was left standing, and it hit the power line. Big sag down in that line. The company had a man out there, they was gonna have a trial shooting juice through it, and they had a man out there for three days. When that happened, that was just terrible! Apt to get electrocuted and the likes of that, with this thing. They came up there, the whole crew to pull the slack out of it. They would just double the slack back between the insulators at the next pole. They wrote me a letter and told me that if any more damage came to their line from my negligence, I'd have to pay for it but they'd stand it this time.

The power company was stringing the high-tension power line from Melwin Dam, about five miles east on the Lewis River. A concrete dam completed in 1931, it was 313 feet high, 738 feet long with a powerhouse capable of 136 megawatts of electricity. The lines were to go another fifteen miles over the hill to Longview, and eventually connect to the extensive Northwest power grid being developed under the New Deal.

That was in 1937. The 320 acres of timberland on the Little Kalama River, purchased in 1930 for $2,500, supported our family and two or three other families throughout the entire Depression. It took seven years to log that parcel (and an adjoining 160 acres), and manufacture the timber into railroad ties, bridge plank, and mine timbers as the market came and went. My sister started school at the one-room Little Kalama school. I was born there in January, 1934, and my brother was born there two years later. It is the site of my earliest memories and might be considered our family origin story.

On to Oregon: 1937

The timber up on the other end of the half section, I hadn't logged. There was lots of alder in there and some bare spots, so I took the other timber off the hill. That was about the time Robert Burr and Dude Ogle had got another patch up the Columbia next to Stevenson [Skamania County, Washington]. I sold the donkey and the cows and what stuff I had [in the Little Kalama] and took the mill to Stevenson to saw Robert's stuff out. Just about the time I had his stuff sawed out, I heard about the Coos Bay [Oregon] timber opening up for gyppos. Tom Powers had been here [at Coos Bay]. He came back up hunting at Stevenson [and told me about it]. We came down here [Coos Bay] and I looked at the timber and made a deal for four forties on the hill at Hauser. So then I went back up [to Washington]. I had three carloads of ties to load. I left Bud Huff and John Linden to load them, and came back here.

ENDNOTES

[1] E. B. Mittleman, "The Gyppo System," The Journal of Political Economy, v. 31, no. 6 [Dec. 1923], University of Chicago Press.

[2]. The gyppo subculture was immortalized by the fictional Stamper family of Ken Kesey's well-known 1964 novel, *Sometimes a Great Notion.*

[3] Parbuckle: A double sling used to roll the logs out of the river onto the bank, using the power of the steam loading donkey.

[4] "China Tiem Producer," Timberman, Sep. 1933, p. 66; "China Building Railroad and Highways, Timberman, Dec, 1934, p. 87. "Tie Mill Aid to Labor," Timberman, Nov 1933, p. 60. A. W. Fairhurst addressed the Business Men's Club and Kalama on November 15, 1933 pointing out that Fairhurst Lumber Company has been marketing a large quantity of ties in China and the small tie mills in Southwest Washington have been a decided factor in relieving the unemployment problem.

[5] Timberman, Sep 1930, p. 156. "Tie Making at Morton: Interesting Small Mill Center."

[6] Paul Lueth, "The Small Moveable Mill," Timberman, July 1937, p. 5ff.

[7] "The West Half of Section 27, Township 6 North of Range 1 East of the Willamette Meridian, containing 320 acres, Situate in the County of Cowlitz State of Washington." It was purchased by George and Doris Youst from Acme Coat Hanger Comapny, L. L. Dillon, President, July 1, 1930. The abstract of the title, in my possession, is a most interesting document, showing that the property had been part of the Northern Pacific Railroad land grant, had been sold and re-sold, and in 1926 sold at a Sheriff's sale to the Woodland State Bank and in 1927 it was mortaged by the Acme Coat Hangar Company to the bank for $2,140. The Youst mortage to the Security State Bank in Woodland, C. A. Button, Pres., was for $2,000.

[8] OWR & N: Oregon-Washington Railway and Navigation Co., the forerunner of the Union Pacific in the Northwest.

[9] Timberman, Nov. 1939, p. 67, obituary.

[10] Bill Yost was one of my dad's younger brothers. Swede Yost was the other (he was not a Swede. It was a nickname which stuck.) Both spelled their names Yost, whereas my dad and grandad spelled it Youst. Family genealogists have discovered that the spelling could go either way within various lines of the family. In our line, it was spelled Youst more often than not.

[11] See OHQ 87 (1988), 5-29. "Agitate, Educate, Organize," by William Biglow and Norman Diamond. See also ILWU Seattle Local 19 website.

[12] Louie Skinner was husband of my mother's step-sister Mabel Godfrey.

George Youst Gyppo logging and sawmills
Coos County, Oregon, 1937-1955

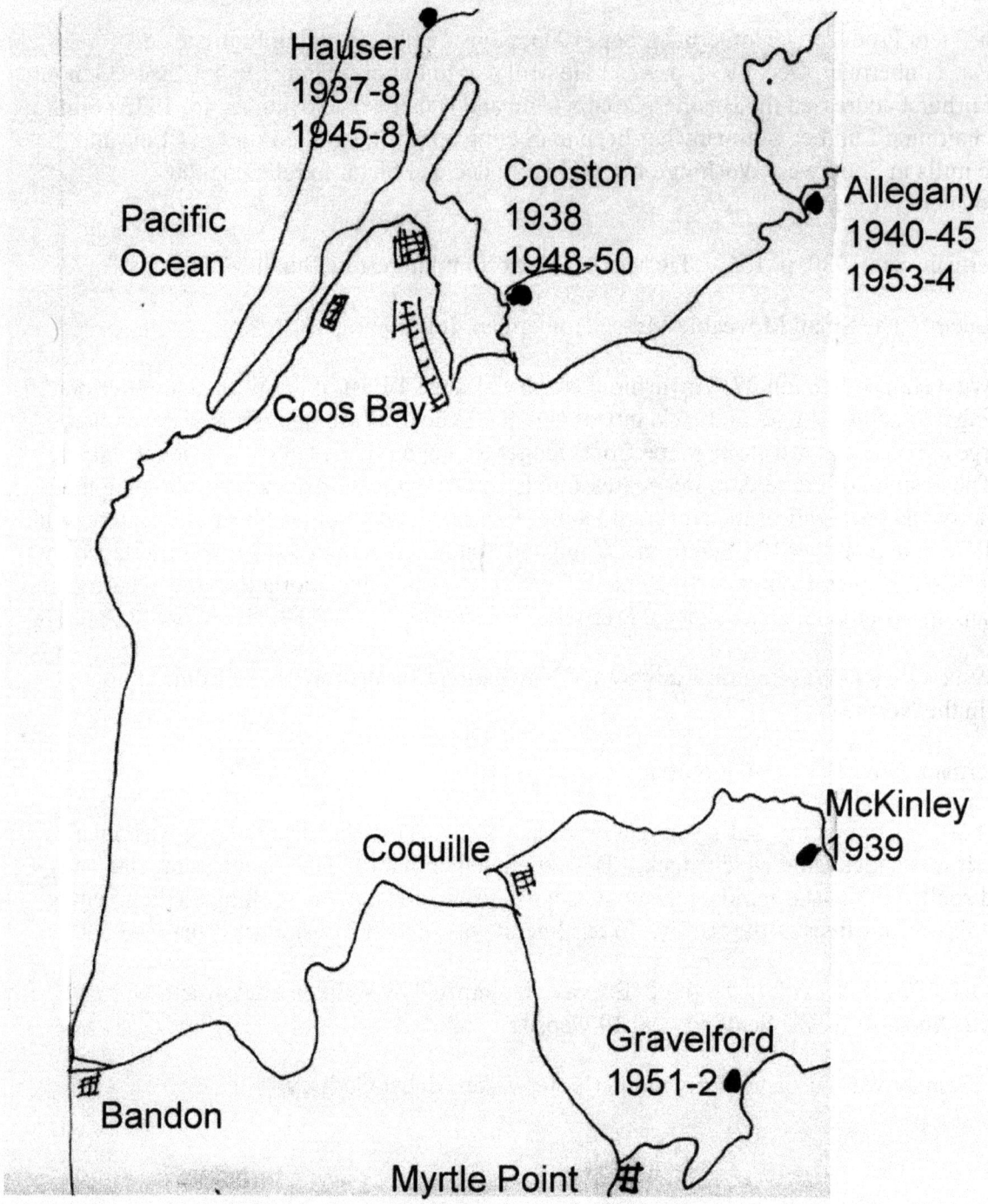

Hauser
1937-8
1945-8

Cooston
1938
1948-50

Allegany
1940-45
1953-4

Pacific
Ocean

Coos Bay

McKinley
1939

Coquille

Gravelford
1951-2

Bandon

Myrtle Point

COOS COUNTY, OREGON

and the

EMERGANCE of the GYPPO MILL

1937-1957

The move to Coos Bay: 1937-38

The move to Oregon was the pivitol event in the history of our family. Dad was 38 years old in 1937, with three young kids and no obvious prospects remaining in the state of Washington. The suitable tie-mill timber was exhausted as far as he was concerned and he needed new land, a place where there was still a chance to do what the Great Depression had taught him to do very well: run a small sawmill business on a shoe-string. Coos County, Oregon, was ripe for the taking. The re-growth from the fires associated with the Coos Bay fire of 1868 was about 65 years old and about right for small sawmills such as his. Around the bay there were thousands of acres of such timber in private ownership, and there were probably less than ten small sawmills in the entire county at that time. What was needed was a timber broker who could could encourage small sawmill operators into the area and organize them, such as Fairhurst Lumber Company of Tacoma had done in southwestern Washington beginning in the 1920's.

Herbert Busterud (1893-1947), president of the newly created Busterud Lumber Co. of Coos Bay, was attempting to do just that. Dad got word while he was cutting ties at Stevenson, Washington, that the timber in Coos County would be opening up for operators such as himself. He came to Coos Bay and looked at a 160 acre patch of 65 year-old "third growth" timber that Busterud had options on. It was located just off Highway 101 about 5 miles north of North Bend, a very desirable location with easy access and a short truck haul for the lumber.

Sixty to 80-year-old timber in Washington was called "second growth." For some reason the same-aged timber at Coos Bay was universally known as "third growth." Which "growth" it was, of course, depended upon where you started counting. There had been about 50 generations of Douglas fir growth and re-growth since the last ice-age! But the loggers and sawmillers in Coos County insisted that the young merchantable timber around the bay was "third growth," and so it was.

It was the timber brokers who made it possible for these small sawmills to operate. Busterud was the the first and only broker for the Coos County area. The broker took large orders and "brokered" them out among the various mills. Herbert Busterud's problem was that there were not enough small mills in Coos County to make it profitable and that is why he was encouraging tie-mill operators from Washington to come to Coos Bay. By 1938 there were ten small sawmills signed up with him, each cutting about 10,000 board feet perday – which gave an aggragate total of 100,00 board feet, the equivalent of the output of a moderately sized large sawmill.

Dad tells of our first days in Coos County,

I met up with Jim Siestreem while I was looking for a house [at Hauser, near Coos Bay].[1] He had an old truck – he'd run the front wheel into the ditch, and I asked him if he knew where I could find a vacant house. He says, "If you'll help me get this truck out, I'll go with you and see what we can do." He took me up to Ernest Peterson's house--the one we moved into.[2] There was some people living in it, but we went up to see Ernest. Ernest says, "My cousin is in there now. I couldn't very well kick him out." Jim says, "You better let this fellow have it. You won't get no money out of your cousin." Ernest got him out, and that's when we moved in.

The room off the back porch was full of papers and boxes and things and we had taken them out and was burning them in the yard. All at once an explosion –

----

[1] Jim Siestreem (1891-1963) was the elected Chairman of the Coos Indian Tribal Council at that time. He became a lifelong friend of the family.

[2] Ernest Peterson (1894-1955) had a dairy farm on North Inlet, Coos County, Oregon. There was a samll patch of old-growth Douglas fir on his place, which Dad logged in 1938. Ernest Peterson had his own "gyppo" sawmill during the early 1950's.

knocked Laurence[3] down – pieces of metal in his face – still has some in his ear. I got a dynamite cap down my leg, I still got that too – and pieces across my belly. Cut my lip. Evidently a box of dynamite caps in that paper. She blew up on us. That wasn't a very good start.

I remember family discussions over the possibility that the dynamite caps had been planted by the previous tennants, in retaliation for having been evicted in favor of paying renters. That thesis was rejected, however, and it was generally agreed that the caps just happened to be there and it was mere bad luck that caused them to be thrown into the fire. The casual attitude toward dynamite among loggers in those days make it a wonder that there were not many more accidents than there were. In any case, the explosion was a disappointing start for life in Oregon.

Jim Siestreem came up with his [mule]team and helped get the foundation down for the mill. He thought the Indians needed a team of mules, so he went up to Fort Lewis and talked them into giving him a team for the Indians. They [the Army] was getting rid of mules anyhow, so they gave him a beautiful team. They had Uncle Sam's name [US] right on 'em, and old Jim, he was stuck with the mules! He was [also] instrumental in getting this Indian building out by Empire for the Indians.

George Youst's first sawmill in Coos County, 1937. Located near Hauser, it is set "in the woods." Note the 65-year-old Douglas fir timber still standing behind the mill and the stumps on each side of the plank road. The truckload of lumber, ready to go to town, contains 100 16 ft 2x10's and 12 24 ft 2x10's – about 3,100 board feet of lumber. This mill might put out three such truckloads, on a good day.

---

[3] Laurence Youst, my brother, b. December 1935.

When we first set up the mill, a guy with a little bitty donkey yarded for me. Five-eights line for a main-line. He couldn't do much. Had to buck the logs. Couldn't pull in the whole tree with it. We cut off one setting with that.

This was in the days before power saws and the small sawmills usually logged directly to the mill. The logs tended to be tree-length and would be bucked by a gasoline dragsaw at the mill, thus avoiding the time and labor that would have been needed to buck them by hand in the woods. Dad had sold his 10 by 11 steam donkey in Washington before we moved to Oregon, expecting to use gyppo loggers for his mill here – at least initially. The very small donkey with its minuscule main-line couldn't handle tree-length logs, and so he soon had a donkey of his own and from then on always did his own logging, as he had done in Washington.

Then we moved to another setting. That's when I bought that donkey that had burned in the Bandon fire [of September, 1936]. Bud [Huff] and I built a sled for it, babbited it up. It was a Willamette. We had to go to Portland to have the friction blocks made for it.

It's amazing, but a steam donkey could actually survive a forest fire. The bearings were babbit, and new babbit could easily be poured in the woods. The new sled, of course, was built in the woods. The friction blocks were of hardwood, and had to be re-manufactured to the specifications of the original blueprints at the Willamette Iron Works plant in Portland. The donkey was a double-drum Willamette 9 by 10, serial number 67. According the Merv Johnson, currator of the Willamette Iron Works records, it was built in Portland in 1903 and first sold to Siulsaw Lumber Company of Gardner, Oregon. It evidently had moved from one logging company to another in Southwestern Oregon until it was unlucky enough to be in the path of the devastating Bandon fire of September, 1936. The Coos Bay Iron Works at that time advertised "steam to gas conversions," and Dad removed the firebox, boiler, and steam pistons and had Coos Bay Iron Works install a new 85 hp Ford V-8 engine on the old steam donkey. It proved to be a fine machine for its purpose and he used it until at least 1947.

Dad had gained early experience in salvaging steam donkies lost in forest fires. In 1914 his first job in the woods was helping his dad (my grandfather Frank Youst, a "donkey doctor") with a contract to build new sleds for five steam donkeys that had burned at McCormick Lumber Compay in Lewis County, Washington. That early experience proved to be of great value to my dad during all his years in the sawmill business. He built any number of donkey sleds of his own, and the consruction techniques used in donkey sleds transferred very well into building solid sawmills. "Hell for Stout" was my dad's motto, and whatever he built would stand all the cosmic forces of logs and machinery.

> **We rigged the tree. Bud [Huff, 18 years old] was in the tree, hanging the guy-lines, and he dropped a clevis pin down in the brush. Old [Herbert] Busterud, the broker, was there. We hunted and hunted, and we never found it while he was there. Afterward, Busterud asked me, "Did you ever find that clevis pin?" I said, "No, we never found it."**

This scene is very representative of the popular conception of the shoestring operation of a typical "gyppo" of that time. Family crewmembers, in this case Dad's 18-year-old nephew (my cousin) Bud Huff (1919-1976), doing the high-climbing. Then, dropping the clevis-pin and not being able to find it, the whole crew stopped to look for it. The timber broker looking on, gaining his first impressions of this sawmiller that he had encouraged to come down from Washington. One wonders what discouraging thoughts might have gone through Mr. Busterud's head as he watched an entire sawmill and logging crew hunt futilely through the brush, looking for a 75-cent clevis pin – and shutting the operation down because they couldn't find it! The next day, with a new clevis pin, the tree finally got rigged and the mill was able to start cutting lumber once more.

> **We was doing pretty good on that setting. Finally we were burning slabs. One night we was in bed, and Ernest Peterson had been to the show in town. He was falling timber for us. He came by and woke us up. He says, "Your mill's burned up. If you get up there right now you might save your donkey. The mill's gone." So we beat it up there. There was fire all over the sled when we got there. We beat the fire off the sled and saved the donkey. That's all we did save. Then I pulled the donkey up on some sticks [small logs] to load it up. I figured I'd have to go loggin'.**

If the explosion of the dynamite caps on the first day appeared as a bad omen towards life in Oregon, the burning of the mill in June, 1938, would seem to be a last straw. It did not, however, slow my dad down one bit. The next morning he was at the office of the Irwin and Lyons Lumber Company Mill-B in North Bend to ask if they had need of a logger with his own yarder. He overheard Jim Lyons (1903-1953), the owner, whisper to the bookkeeper, saying, "That's George Youst. His mill burned down last night. Give him anything he wants." As it turns out, a gyppo logger selling logs to Irwin and Lyons was John Aason, and it just so happened that he needed a yarder at that time.

John Aason----he died. It was in the paper last night--85 years old. John had timber at Willanch. He wanted me to come up and yard it. I didn't know John, so I made up a deal with him--so much a thousand to yard it. We took the donkey over there--rigged the tree and started logging. Boy oh boy--good show. Got the logs in fast. John had a couple guys hauling for him, and another donkey there, loading.

George Youst (facing front), gyppo logging after his mill burned in 1938.

The old bugger, you know, would scare you do death. He'd get out on the pile. All they had was a quarter inch straw line for a haul back--skin it back to the spar tree. They was loading like that. Little tongs, and they would pull out, and they'd just be snappin' like that (snap, snap, snap). John would be up there and they'd be snappin' round his head, but they wouldn't hit him.

Not much before that, John had been in a car wreck and they'd sued him--got a big award. But of course John didn't pay it. He didn't have nothing to pay it with. They took his driver's license too. He couldn't drive, so he was in a heck of a shape.

John Aason (1882-1967) was the most colorful of all the gyppo operators ever in Coos County, and a legend in his own time. His parents, from Norway, settled in the Coquille Valley in 1872. There were five boys and one girl. By 1910, the brothers had a fairly large logging operation on the Coquille River – several splash dams and a large logging camp. Later they had their own railroad. The panic of 1929 put an end to the prosperity of John Aason's operation, however, and he fell on hard times. He became infamous among working loggers due to his reputation of sometimes paying his crew with"rubber" checks, checks with nothing in the bank to back them. By 1938, when my dad worked with him, John must have been at the very nadir of his career. He was, however, the eternal optimist. He went to Arcata, California, at the beginning of a sawmill boom there in the late 1940's. He was well into his 70's when he finally made it big again. He died quite wealthy at Eureka, California, in November, 1967, just a few days before I conducted this interview with my dad.

When we got the first raft out, we took it down to Mill-B. Jim Lyons was buying it. The office was full of people with claims against John's money--what he had coming out of this little raft of logs. They wasn't paying much. I think John was only getting six dollars or six and a half per thousand for the logs. I remember the raft only came to $660, about 100,000 feet in the raft.

When the bookkeeper came out with the books, he had all these slips against John. He threw them down on the table and said, "John, I'm no Houdini. How do you want this divided up?" John said, "See that Youst gets his money. The rest of them can wait and get it out of the next raft." This woman says, "If I don't get mine, there'll be no next raft." She was the one who owned the timber![4] So I was scared of John then. Boy, with that kind of a deal, and I'm not making much money anyhow. So I decided to pull my riggin' out of there.

Going out the next morning Bud [Huff] and I decided to pull out. I had enough logged for another raft. So I met old John. He was driving his car--he wasn't supposed to, but he was driving it. I told him I was going up to pull my donkey out. He says, "Don't do that. Don't do that." I said, "I'm scared of you, John." He says, "Why? Why?" I said, "Jesus, all those bills against you. I'm afraid I won't get my money." He says, "When I get through with those fellows it'll look entirely different! And furthermore, I'm dickering for some timber on down the coast that's going to be a money-maker! You just stay with me till we get down there, we'll really make some money." "No, John," I said. "I can't do it." So I went up and moved the donkey out.

---

[4] I told this story, as I remembered it, at a talk I did at the Coos County Historical Association in October, 1982; Dow Beckham included it, with my permission, in his excellent book about splash dam logging, Swift Flows the River, Coos Bay, 1990, page 86.

One-log load, old-growth Douglas fir from the Ernest Peterson farm on North Inlet at the log dump at Hauser, 1938.

Kenny Landers, gyppo log trucker, hauling logs for George Youst, gyppo logger, at the log dump at Hauser.

Then I went up North Inlet and got some timber. It was old growth. Ernest fell it for me. I made a deal for nine dollars per thousand, in the water [tide water, at Hauser]. Had a tree--big fir tree at the side of the road. We didn't top it. Just limbed it up for 100 feet and put guy lines on it. Just short logging, a few trees scattered around. These guy lines that I had--I put four guy lines on it--all I had, on it. And I had them right straight across from each other, all four of them. And we were logging in a bite, a square lead.

There was a fellow living above there, a glass blower [cutter] from Chicago. . . . This one day we was yarding in the bite. Big logs--they was old growth. This fellow had brought two buckets of plums down. He was starting back by the landing and he stood, watching. We were pulling down a lot of alders and things. Well, it was something to see, you know; mashed everything down coming through. So he came back over and stood by the donkey. You was standing there too. By golly, when that guy line broke--when it broke, the two across this way broke. Three guy lines broke at once, right at the tree.

The tree came over, and he ran right square underneath it. You would have been too, if you could have kept up with him. You would have got the tree on you, too. Then we had to rig another tree. Yup, that's the way it was.

I was recording this interview with my dad in November, 1967, and he was speaking to me. It is true that I was present when that spar tree came over and killed Leonard Seaburg. It was September 8, 1938, and I was 5 1/2 years old. I was standing alongside the donkey when Mr.

Seaburg stopped by to watch the logging. When the guylines broke and the tree started over, he ran out away from the donkey, directly under the falling tree and I followed him. We would have been all right where we were alongside the donkey, but Mr. Seaburg panicked and ran under the tree. After we passed under the path of the falling tree, he turned back toward the donkey and was killed when the tree fell directly on him, smashing him into the ground. I could not keep up with him, and so I was on the other side of the tree when it landed. My mother was punking whistle and had been simultaneously baby-sitting myself and by little brother – typical gyppo operation. I remember that she was pretty relieved to see that I was all right. I watched them dig Seaburg's body out from under that big old growth, and they never let me hang around where they were logging after that. Not until I was 15, when I started punking whistle myself.

Dad had used those guylines while logging the small third growth timber for John Aason, and there had been no problem. Logging old growth was quite another thing. He said he was logging in a "square lead," which is to say that the logs are coming in at an angle of 90 degrees from the spar tree, and all the force was against the two guylines that were opposite the direction of pull. When one of them broke, the tree, which was a large, old growth Douglas fir complete with crown, started over and broke the other two guylines on that side of the tree. The pictures on the previous page give an indication of the size of the tree. It was a freakish accident, almost unheard of, but logging accidents, then or now, are not uncommon and are frequently freakish.[5]

Eighty years passed and I didn't know any more about the accident than I have written above when, during July 2020, I received emails from Nancy Benson of Falls Church, Virginia. She said that Mr. Leonard Seaburg was her grandfather. She was doing genealogy on her family and was interested in more information about that logging accident. As it turns out, I am the only living person who knows anything about it and I answered all her questions as best I could. She filled me in on details of who Mr. Seaburg was and a little bit about the family, all things I never knew. He had apparently moved with his family to Oregon from Chicago, where he had worked as a glass cutter. They were living at Hauser, on North Slough, probably near the logging that Dad was doing at that time. Mr. Seaburg was a Swede and was apparently a close friend of another Swede, Victor Enlund, who lived near there, and that's where the two buckets of plums came from. Mr. Seaburg was bringing the ripe plumbs as a gift to share with dad and his crew and ended up in the wrong place at the wrong time. A sad story.

---

[5] I had covered some of the foregoing material in an interview with William G. Robbins in 1984. He mentioned part of it in his book, Hard Times in Paradise, Seattle, 1988, p. 113.

When we got through with that, I hunted for timber for a long time. Couldn't find anything that suited us. You know, they was putting specifications on the lumber. They wouldn't take that coarse grained lumber you get out of this second growth. I came pretty near taking a mill Busterud had an interest in, then he had one he wanted me to put on the dock. [This was the Portland Dock, next to the big Coos Bay Lumber Company sawmill].

I can remember my dad and my cousin Bud Huff going out day after day looking for timber during the winter of 1938-9. The way the small sawmill business worked in those days was that first you had to find a suitable patch of timber – Dad wouldn't consider less than 160 acres (four 40's), but it had to be of high quality because the grading specifications had become more restrictive. There was, for example, a requirement that there be not less than 6 growth rings per inch, comparatively slow growing timber. Most of the timber close in by the bay was young and very fast growing – as little as one or two growth rings per inch. The market would no longer accomodate such low quality lumber. Once having found the timber, it was a comparatively easy matter to leverage it into whatever loans would be necessary to set up a mill. The fact that Dad's mill had burned the year before was apparently not a serious obstacle. He was a sawmill man, and he had connections with the broker Herb Busterud, and with Jim Lyons of the Irwin Lyons Lumber Comapny, and others and they would steer him to one available patch of timber after another. There was something wrong with all of them until finally, to put food on the table, he got a contract to do the logging for a mill on Cherry Creek, a tributary of the North Fork of the Coquille River (post office address: McKinley). Dad's one remaining asset was his logging donkey, the one he had salvaged from the Bandon fire.

The Levison sawmill on Cherry Creek was an interesting variation on the small sawmill phenomenon. Until the nearby timber was logged off, logging was done directly to the mill but it was a much larger mill than was customary for that arrangement. They cut as much as 30,000 board feet per day and all of the lumber went directly to the Moore Mill and Lumber Company sawmill in Bandon. I don't know what arrangements Levison had with Moore Mill. Moore Mill probably owned the timber, and they may have had a dominant financial interest in Levison's mill. Both arrangements were fairly common, but the lumber production statistics do not mention the Levison mill – all of its output was included as part of the Moore Mill production. In general, the statistics for lumber production tended to ignore the production of the small sawmills, incorporating their production into the production of the larger mills that took delivery of the lumber, and thus the documented significance of the small sawmills is historically understated.

I logged for Dave Levison that winter, up at McKinley. Five dollars per thousand for putting logs in the water--falling and bucking and yarding with an 11 by 13 Willamette steam donkey. That was quite a deal too. He had an 11 by 13 Willamette steam donkey he'd been logging with. He had Davey Dugger, an old Superintendent for Coos Bay Lumber Company, a good logger, but he'd got all he could get of the timber Levison had with this big donkey. He'd tightlined down the hill, and had an awful time trying to get them in the water. Although he had this big 11 by 13 Willamette steam donkey he couldn't get pressure enough to lift those logs.

So this show I had was at least 500 feet further to tightline than he had. I had to yard them and tightline them into the water. Old Davey Dugger says, "This I want to see." He didn't understand tightlining like I did, because I had put in lots of wood logs. I understood that the further you get away from the spar tree the less power you've got, and you just can't lift them when you get way out there. It's like being at the end of a pole. Davey didn't understand that.

Levison had built a dam on Cherry Creek which impounded a fairly large body of water for the mill pond. The logs were "tightlined" by gravity directly off the hill into the pond. The drawing at right indicates the gravity "tightline" method that was used. The steam donkey was at the bottom of the hill and it raised and lowered the 1700 foot-long skyline. Dad's yarder was at the top and while they were swinging logs into the pond, it would "haul" the carriage back up the hill for the next turn. According to Dad, It only took 30 seconds to descend the 1700 feet from top of the hill to the mill pond with two large logs.

So we moved our donkey out to the pond [to move it across, and up the hill on the other side]. There wasn't a log in the water. Levison said we could get a few logs up at the old setting--they was trucking in there. He went up and got only four logs, brought them down, and dumped them in the pond. I said, "That's not enough to go across that river on." "Oh yes, they'll hold you all right," he says.

I pulled them logs up--Vic lined them up in front of the sled. We got a hold across on a stump, on the other side of the pond and started pulling on it. I was running the donkey, but I wasn't so smart. As we moved out into the river she was doing OK. All the weight is on the back end of the sled, and as long as that was on the bank, I was fine. Those logs stayed out there and held the front end up, but just as soon as I came down off the bank, those logs rolled right up to the front end and the back end went right down under the water. Well, I just kept going until the engine went under water and she quit. I climbed up on top of the haulback drum by that time, to keep out of the water. There I was, in the river, with just the nose of the donkey sticking out of the water, and on top of the haulback drum! I had to fish a log out from underneath and ride it back to shore.

I didn't know what to do for a little bit. They had this old 11 by 13 setting there. I told Levison, "How about firing that steam donkey up and pulling this thing out." "Anything you want," he said. So we fired it up and pulled our donkey back up on the bank on the other side. Drained the pan and oil and everything. Cleaned her out. Had to bring out new oil, for the transmission and everything. Got her together, and the next day we took 'er up the hill.

The next page shows a steam donkey crossing a mill pond on its own power (from Willamette 1925 Catalog, page 43). After the steam donkey went under the water and the fire was extinguished, there was enough steam pressure remaining for the donkey to pull itself the rest of the way across the pond. There was 150 pounds of steam when they started and they still had 75 pounds of steam after they got across the pond. My dad's gasoline donkey did not have that advantage. When the Ford V8 engine went under water, that was the end of the power and there it stopped!

12 x 14 inch Humboldt Yarder about to ford a slough 160 feet wide, 8 feet deep, 150 pounds steam pressure.

In 8 feet of water. Fire entirely extinguished. Engineer still at his post.

Emerging on the opposite bank. Steam pressure 75 pounds.

Dad's 9 by 10 Willamette steam donkey, s/n 67, converted to an 85 hp Ford V8. Note the large bull gear (right), and the small core of the mainline drum. Geared down, it was a very powerful donkey.

We had to pack gas on our back. I had Vic Graham and Bud Huff, and me. Two of us would take a five gallon can of gas up each morning on our back. That's what we run on. We got that tree rigged up above, and started yarding. Those first logs that was up there had been bucked for that 11 by13 Willamette. Big, long logs. 64 feet long, some of them. I was surprised my little donkey pulled them in. Just an 85 horsepower V-8 Ford. But she was geared down, and that big old bull wheel on it had a lot of leverage, and a small core on the drum.

We logged [the long logs] in, and then we didn't have enough line to go down the 1700 feet to the river. Julius Benham [of Benham & Laird Logging Co.] had a skyline he wasn't using, so we went over and rented it from him. We got the carriage from him, too. We strung the skyline down there, and took the old Willamette and put a block on the end of the skyline, tail holded it, and put a block in the tree. We had an extension on the end of the skyline because it wasn't long enough. It just did reach from the tree to the water.

We'd hook on two big logs, and Vic was about halfway down the hill giving hand signals with a flag. We'd go ahead on that steam donkey, lift 'er up and all I had to do was keep the haulback from getting tangled up, she went down so fast. 30 seconds from when she left the tree till she was in the water. 1700 feet. She was getting down there pretty fast. When she hit the water, the steam would just fly off that carriage, it developed so much heat going down the hill. I used my haulback to pull the carriage back. He [the steam donkey puncher] would raise the line up a bit so I could skin 'er back, then he'd slack'er, and I'd have to pull back slack enough to get onto the pile [of logs, the cold deck at the top of the hill]. That's the way we got'em down there. Yuh!

Bud and I and Vic was boarding there. Poorest boarding I ever got – those Advents. They had baloney for breakfast. Oh, I tell you, if you really want to loose weight, just go boarding with some Advents. I'll guarantee you. They're better cooks than the English, but it's *what* they cook! But I'd say they're *better* cooks. There's nothing as bad as the English for cooking, in my notion.

Willamette 11 by 13 steam donkey working in the woods. Photo is from the 1925 Willamette catalog.

Dave Levison was a Seventh Day Adventist, one of many small sawmill operators from that faith throughout the west. There were two principle differences between an Adventist sawmill and any other one: First, they never worked on Saturday and second, their dietary customs were often unlike what most loggers and sawmill workers expected. No one ever complained about not working on Saturday, but old time loggers such as my dad expected a cookhouse to serve the kind of food they were used to – hotcakes and bacon and eggs and fried potatoes for breakfast, and lots of them. Nothing like that was served at the cookhouse of the Levison mill!

From 1935 until 1940 there had been an Adventist sawmill near Brookings, Oregon – a few miles north of the California line. Dave Levison may have been from that sawmilling community, I don't know. But Vic Graham, an 18-year-old youth was from Brookings, he was from a Seventh Day Adventist family, and he was at Levison's mill when my dad came there to log in the fall of 1939. Vic went to work for my dad and stayed with him almost the rest of his life. By what may have been a coincidence but was possibly more sinister, Levison's mill at McKinley burned down in June, 1940, and the Brookings Adventist mill burned down a few days later.

When we got through up there, we moved [the donkey] down the hill. Got down to the pond and everybody in the area was there. They'd heard I was going to take the donkey across the pond again. And the river was up. They'd heard about me gettin' 'er sunk the last time, so they was all setting up in the mill, watching. Raining like heck. The river was so swift. They had a boom across above, but we had an awful time getting any logs loose, to get logs to make a raft to bring the donkey across on. The pond was pretty wide.

59

When got over to the other side, the bank was straight up. I'd have to jerk it to get it up, and every time I'd jerk it, another one of them logs would go out from under the back end. I wasn't real sure I'd get out of there without having to steam up the old 11 by 13. But I finally got it up on the bank. Leveled 'er up and that's where we left it.

Levison had promised me another forty to log. But the day we got the donkey up on the bank, a guy came over that had been logging with a Cat. He agreed to build a landing and rig a loading tree and yard for less than what I was going to do it for. I was getting $5.00 per thousand. I could have made some money on that show for $5.00, but that guy didn't make any money with his cat. He got the job, but Levison told me he had already put in a bid for government timber [BLM land near there]. He'd give me the job yarding that. I took the job loading those logs, that this guy yarded with the cat. We loaded them out, and in the meantime I'd got hold of Bob Wilkinson and the timber above the falls.

Dad, center. Loading logs for the Levison mill, winter 1939-40. Note plank road.

Gyppo Sawmilling Above the Falls
An Oral History

Gyppo logging for a small sawmill was never going to generate much more than wages. To make real money, there had to be a "value added" to the product and value is added to logs by making them into lumber. Dad had to find suitable timber for that purpose and he kept in close contact with Herbert Busterud, the broker. One day in early 1940 he was at Busterud's office and Bob Wilkinson was there trying to find a way to convert into cash some very high grade though almost hopelessly remote timber he and his brother Cleland had inherited from their stepfather, Joe Schapers. Cleland, his mother, and his two daughters lived on the place, and Joe Schapers had willed the 160 acre wilderness homestead to the two step-sons and he willed $500 each to the two step-daughters. There were back taxes on the homestead and no money to give the girls their inheritance. Their only possible source of cash was in the timber at the homestead but they could not find a buyer because of its remote and apparently inaccessable location.

The Joe Schapers homestead before Dad's mill logged the timber. The first setting of the mill was in the timber shown on the hill opposite the ranch house. The second setting was along Glenn Creek, to the right of the picture.

Golden Falls, Coos County, Oregon. The Joe Schapers homestead was one mile above the top of the falls.

Dad talked to Bob Wilkinson, went up to the homestead, and looked at the timber. The homestead was in the narrow valley of Glenn Creek, above the Golden Falls about 11 miles beyond the remote settlement of Allegany. A few families had lived in that valley since the 1880's, but no one had imagined that lumber could be economically manufactured up there and hauled the 30 miles to Coos Bay over an almost impassable road. Dad thought he could do it.

I guess I got connected with Bob Wilkinson through Busterud. Bob was trying to get some money out that timber, and he had been to Busderud's. Bob took me up, and I'm telling you she was a pretty rough show. That road was <u>bad</u>. I came back and started telling Doris and Bud how bad it was. I says, "It's up to you guys, if you want to go up there." I took 'em up the next day, and when I got over the Falls they said, "Where's that bad place?" I had made it sound so bad they were expecting worse! So I made a deal with Bob for that. I got the timber for a dollar and a half a thousand for the first forty.

I had to take the donkey off the sled because we couldn't take it over that hill [the Golden and Silver Falls road was too steep, narrow, and crooked]. Bud and I had to build a new sled up there. Had to fall some trees. Got the donkey up there, and unloaded it on the ground. We tied it to a stump and yarded in a couple sticks to make sled runners. We got the sled built -- a good sled, too. Then we moved it up the hill.

The mill we got had been at Tenmile Lake. A real estate guy had bought the mill and had it in his back yard. Bob Wilkinson had made him a deal for it. It had a long carriage -- 24 feet long. I cut it down, made it shorter, lightened it up a quite a bit. The feed works was no good at all. We used it on the first setting. The flywheel turned, then there was a paper wheel you shoved in between two wheels. If the carriage didn't start just as soon as you hit it, you'd wear a flat spot on the paper. Many times I'd have to take it to town to get new paper put in it or get it turned down with the lathe to get it evened up again.

The Willamette 9 by 10 yarder with its 85 HP Ford V8 engine and the new sled Dad and Bud Huff built for it above the falls.

The carriage feedworks of a sawmill allows the sawyer to control the speed that the log engages the saw and thus is of crucial importance for both the quality and the quantity of lumber that can be produced. The original water-powered slash mills of long ago used a rack and pinion to move the carriage forward, but that system had been discontinued except in some small, portable sawmills. Very large sawmills tended to use a steam piston feed, which was fast and efficient. But the large majority of sawmills by 1940 were using either the variable friction feedworks that Dad was complaining about, or the more reliable variable belt feedworks that he installed later, during World War II.

The completed mill, June, 1940, set in the timber at top of a hill with no road to it.

We set the mill up after we got our donkey sled built. We set it up on top of the hill, then we shot the slabs down into the field so we could burn them away from the slash. That worked out pretty good. We let the slabs pile up in the field. There must have been a hundred, hundred and fifty foot jump for the slabs to jump off the end of the chute out into the field. They covered quite an area. Then, Christmas eve, they set it afire and it lit up the whole valley. For a long time it burned.

Slab pile with slab chute coming down from the mill at the top of the hill. The lumber chute was several hundred feet further down the field, to the left.

I figured I'd have to chute the lumber down the hill. I didn't know just how it was going to work. I had Bud and Vik and Fred Debuque dig a hole by the county road to make a landing for the lumber. When we got the mill running, and started chuting the lumber down, heck, it wouldn't even slow down when it hit that little platform. I had to build the chute clear across the field. Then had to make a flat place for it, must have been a hundred feet long, to slow it down. Then we put a big log at the end with a piece of plate steel to keep from beating the log to pieces. And it did beat two pieces of steel to pieces. We had tires, too, to cushion it. The hemlock would go down so fast they would just accordion for three or four feet of the end of them when they hit.

64

Pictures of the mill, first setting, above the Golden Falls at Allegany, Oregon, 1940

Crew eating lunch on the carriage.

Dad running the drag saw, cutting the log into 24 foot lengths for the mill.

Mill under construction, showing carriage tracks, saw husk, and part of log rollway. Dad hand hewed the carriage track stringers from logs.

I tried to cut orders when we started, but those big timbers, when they went down, they split that big log in two. Drove right through the big log! So then I got down to where I just cut 2 by 4's, that's all. We could handle them. They were all the same size. But even those sap pieces of 2 by 4's, they'd go down and jump all over the place!

Then the donkey engine quit. We had to take it out and take it to town, get it fixed up, and bring it back. Bud and I had to pack that motor back up the hill. I'm telling you! We had a pole on each side -- one in front and one behind, packing that motor up the hill. Just the two of us packed it up there and put it back in. It worked all right after that.

The "Above the Falls" Road

But then, to get somebody to haul [the lumber], that was the next thing. The first three or four guys that we got to come up to haul — just one load and they wouldn't come back. Finally Sedgy, he had a truck. He took two loads out. He said, "You just as well get out of there." John Aason told me so too. He said, "You just as well get out of there. You'll never be able to get any timber down over that road." That's when I said, "I think I know a guy who can get it out."

I just sent a "mental telepathy" to Dude, up at Stevenson. I never wrote him a letter, or nothing. Sent him a "mental telepathy" and in three days Dude was down here. He says, "What do you need?" I says, "I need somebody to haul this lumber out of here." He says, "You got anything to haul it on?" I said, "No." He says, "You get me something to haul it on, I'll haul it."

Landers [Kenny and Lionel Landers, of Allegany] had an old V-8 Ford truck that had hauled logs for Sjogren and Whittick until they had wore it out. Then Landers bought it, then I bought it from Landers! I think I paid Landers $600 for that old truck and trailer.

Dude started in on that. He was staying with us. The first day we loaded him

66

up a load of lumber. He started out and didn't get a hundred feet from the dock till he was stuck. We worked on the road that day. We went and cut a bunch of planks to plank the road. So Dude came in that night and he wrote on the calendar, "Stuck".

We got the road planked out to the county road. The next day he got about 300 feet down the road, and stuck again. So that night he came in and wrote on the calendar, "Still stuck!"   The third day, he got to town with it, and back. Every day after that, one day he'd get one load and the next day he'd get two loads. You bet!

Planking the road at Frog Creek, a few hundred yards above the top of Golden Falls.

Roy Strickland, down at the store [at Allegany] called him "Perpetual Motion." Couldn't stop old Dude! Down at the dock, they called him "Old Dirty." Those guys told me that everybody else, when they took their chains off, they'd hold them out away from them when they loaded them into the jockybox, but not Dude! He'd grab them in his arms, right against him [the muddy load chains].

The tires weren't good on that old truck. Dude had a flat tire one night. He didn't get home till pretty late. The only tools he had to change that tire was a bolt and a railroad spike, to take that tire off and fix the tube, and put it back on. A truck tire at that. Yeh! You couldn't stop that fellow. No way of it.

Dude fixed the road up a little bit all the time he was there. He'd see some rocks along the road and he'd pick them up and throw them in the jockey box behind the cab. He carried a sledge hammer with him. Anything he could lift, he put in there. When he'd get down to one of those big mud holes, he'd pack those rocks around to the front of the truck and throw 'em in the hole and take his sledge hammer and beat 'em up. Hard sand stone is what the rocks was. Be golly, by the time the snow got off up at Stevenson, where he was at, we got to where we could hire some truckers.

We got to producing too much lumber for one truck to handle. There wasn't quite enough for two. We had an extra truck, and Doris started taking one load a day. She said if I'd take the load down the hill [over the Golden and Silver Falls road], she'd take it to town. So for several days I'd take a load down and

Mom by her GMC truck with 4,000 board feet of dimensioned lumber on the way to town – a 60 mile round trip and she made two trips a day through most of World War II. Note, she wears a dress, always a lady.

she'd take it on into town and bring the truck back. Finally one day she said she would take it all the way down that day. I said, "I'll follow you." I followed her, and she made it all right. She took it every day after that. And that was that tough show nobody could get out of there with! Yeh!

The Landers started hauling again. The Landers was hauling when they broke that bridge down. Two of Landers trucks -- one was Lionel's truck -- he had another young fellow driving it -- and Kenny was driving the other one. Kenny would make one load and the other guy would make two loads -- three loads a day we was getting out of there. Then, they'd haul over the weekend too, when the mill was shut down.

Anyhow, Kenny was coming back, empty, and the other truck was coming down with a load. He drove out on the old bridge, rotten. He came back to talk to Kenny, and while he was talking to Kenny, two bents of that bridge went down, with the truck. He got in with Kenny and came up to the mill and told me what was the matter.

I loaded a bunch of rigging and blocks and line in the pickup and drove on down. I got down there and seen what was the "diffugulty." The first two bents had gone down. The front wheels of the truck was still on the third bent. It was steep-- the trailer and truck was down like that. I got up on the hind wheel of the truck and took my ax and, whang! I took that chain, and bang! The whole load went clear to the bottom of the canyon just that quick.

The Silver Creek Bridge, built in 1909. In 1943 the first two bents of the bridge – far right – were rotten and collapsed with a truck load of lumber from Dad's mill. He brought the mill crew down, cut logs to replace the rotten posts, cut plank at the mill for the stingers and deck, and put the bridge back into service in a few days. That repair was still good in 1958 when the county finally removed the bridge and closed the road for good.

Just one lick. There was an awful strain on that chain, the way it was laying. Then we pulled the trailer up on the road. I rigged a tightline across to lift the back of the truck up with Kenny's truck. He'd jerk it, and lift the back end up, and we had chain binders and chain on the front and we'd pull that truck ahead what we could till the frame would hit the bent again. Then Kenny would back up, give it another jerk, and that would raise her up a little bit more and we'd pull it ahead and that's how we brought that truck right back up on the bridge. Just bent the running boards a little bit.

Then we went up on the hill and fell a couple trees, drug them down with

Kenny's truck. We rolled them to one side and bucked them in the lengths we wanted for posts. Then Kenny would back up, we would roll them back in the road, and he would push them ahead with the truck. Then we'd stand them up, put the bunks and planks on them, braced her up, and there we put in those two bents. We put stringers on top of them, and we were in business again!

That load of lumber was down in the bottom. I took the tightline and put it over there, and Kenny run his truck. We tightlined the lumber over and put it on the other truck. That worked out pretty good. We didn't loose any.

We had the one truck of our own. I always made it a point if I could hire it hauled cheaper than I could do it ourselves, I'd hire it. If I couldn't, I'd haul it ourselves. A while after that, I had two guys driving. They would take turn about. They would run that truck about around the clock, trying to keep the lumber out. Couldn't hardly do it, though. So I got another truck from Pat Rooney. Got it inspected and sent it up. The next morning I was taking the crew up the hill. Below the big hairpin turn (between the Golden and Silver Falls) I met the old truck coming down. Harry [Harry Able, the driver] says, "Where'd you meet Rex?" I said, "I didn't meet Rex." He says, "He was just ahead of me." I said, "He didn't get down this far." So we started walking back up around the curve, and here come Rex up out of the brush from below that hairpin curve. He just had a little scratch on his forehead, and lost his cap.

They charged the same insurance -- it was half what the insurance on the truck was, for the trailer. I didn't put any insurance on the trailer. The truck was totaled and only the tail light was broke on the trailer! Cost $5.00. That's all it cost. I beat the insurance company that time. I lost my down payment on the truck. All it was insured for was what I owed Pat. Then he let me have another truck -- that GMC. Had to sign a note for the down payment on that. I didn't have money enough for another down payment.

When we got through with that setting, there was some timber too far to get, and too big for us to saw. I decided to haul some of it to town -- put it in the river and splash 'em on down. Sell them as logs. So we rigged a tree and hauled some of them down. The logs the size we could saw, we hauled them over and dumped 'em in our pond.

A load of peeler logs too big for the mill. Dad hauled several loads down over the Golden and Silver falls to dump them into the East Fork of the Millicoma River at the Brady and Neal logging camp. There were splash dams in the river, used to drive logs the ten miles to tide water where they were rafted and towed to Coos Bay. Those were the only logs ever to be hauled over the Golden Falls road.

The Dam

We built a splash dam there in the creek [Glenn Creek] and set the mill up in the field, in the potato patch there. We had built the dam [for the mill-pond] with what you call a dead-head, which held about four feet of water. That gave us water enough to float the logs so we could get them up in the mill. Then we had eight or ten feet above that. We had logs across the top, and we put plank that rested on the lower log. We could open that dam up and let the water through for high water. [The upper eight feet of impounded water could be released.]

Splash dam on Millicoma River West Fork, after a freshet had brought logs down on top of it.

I had a guy living in a shack right by the dam. He had instructions that if the creek started rising, like it was going to go over the top of the dam, to open it up and let the water go through. But we had a big rainstorm. We just got to saw one day, in this new mill, and next morning we went down there and the water had washed out around the end of the dam, and the top log had dropped down, and there she was.

She was a pretty sad looking deal around then. I told the crew, "I've got money to pay you what you've got coming as of today. If you want to stay and help fix this dam up, and wait till we get sawing to get the rest of your money, well, OK. If you don't, well, just go on home and we'll fix 'er up ourselves." Nobody went home. They all stayed.

We rigged up. We had a spar tree across the creek and we put a tail hold on the other hill, so the rigging came right over the dam. Slacked the mainline down and put a choker on the top log, tightlined it up in the air – put the haulback on to hold it so it wouldn't slide endo. We lifted it up and cut a couple posts to set underneath of it, let it down on the posts, and that held it up. There was about 20 feet or so had washed out around the end of the dam. We put in braces back to the bank. Had to put two more logs in, cut planks, and put them in. You could tell when you hit bedrock, going down through that soft muck and stuff. Bud and Dave Browning were in the water up to their chins getting those planks in. We got 'er together; finally got 'er fixed up.

The mill in its second setting. The mill pond with logs in it, left. The dam in Glenn Creek was behind the mill. This picture was taken before the roof had been extended over the loading docks, right. This mill averaged about 20 thousand board feet of dimensioned lumber per day through most of World War II.

We sawed out plank to plank the road 1000 or 1500 feet through Cle's [Cle Wilkinson] field to get to the county road again. We done pretty good on that setting.

The loading dock area of the mill, showing loading jacks. The trucks would back under the loading jack, which is pre-loaded with 4,000 board feet of lumber ready to go to town.

When we got through with our first spar tree we moved up on a hill. We yarded and tightlined the logs into the water from there. When we got through with that setting, I gave Roy Mast the contract to log, and he logged the rest of it. We built another splash dam above. We tightlined the logs into the pond that was above the dam.

When they'd get the pond full they would send a guy down to the mill to tell me to open the dam below so the water wouldn't run over the top of that one. When I opened the dam, I'd just let those planks go that I opened to let the water on through. I'd pick up just any old scrap plank to fill in the gap. [This gave the dam a ragged appearance, with planks of random length sticking up along the top.]

One day the bucker -- the one they'd send down to tell me -- he'd got mad and quit. So when he got down to tell me, the water was already going over the top of the dam. He says, "You better get the dam open." I looked out and there it was, going over the top.

I dashed out to open up some hole to let the water through, but there was so much weight against it I couldn't move the planks. There was one plank sticking up above, quite a little ways. I worked on that one, and couldn't get it

73

to raise enough to slip out. Then, like a fool, I put my arm around in front of that plank. And when I raised it up, whang! she took me right over into the water and I came right back through the hole where the plank went out. How I went through where a foot board came out, but I did.

I just curled up like a spider when I went through. I'd seen spiders fall, and they curl up like that. I hit that water--lots of water coming through, at least eight feet of it deep, and a foot long. It made an eddy, and was washed out pretty deep. I remember coming around underneath of that water--it forced me down a third time before I got to the rocks and could pull myself out in water I could stand up in. The whole crew was out there. They thought I was done for by that time. Funny thing, never hurt me a particle. Didn't even get cold, and in the wintertime too. That was quite an experience right there.

The Crew

Men kept getting more scarce. They took the men for the war. It got down so about all I could get to work was old winos and some of those fishermen, and the likes of that. They couldn't pass the examinations. We went on running the mill. I got a rupture and went to get operated on in November. I left Pete Lusignot and his brother Gus [Johnson] up there to fall

The loading dock area with the Ford truck loaded and ready to take 4 thousand board feet of lumber the 30 miles to Coos Bay.

timber while I was in the hospital. They came down Thanksgiving time, and I asked how much money they needed. "Oh, we don't need much." Pete rattled the money in his pocket. I think I gave them $75. Not very long after that, they picked 'em up and put them in jail. Pete and another old guy. They called up the Allegany Store. They wanted me to come down and bail them out. They were in jail over there and wanted $600 to get them out. This old guy, I'd never seen him before. They'd been moonshing all this time! I asked them how much they had coming for the falling. Pete said, "That's all right. That's all right!" I'd sent a bucker out to buck up them logs they'd fell. He came back the same day and said, "What do you want me to do now?" I said, "Buck them logs those fellows fell." Twenty-two trees is all they'd fell.

We had a young feller come to work for me when I was logging for Levison, getting those logs down off the hill [at Cherry Creek, near McKinley]. He was just a kid then, maybe around 18. He just took up with me, I guess. Stayed with me just like my boy! It must have been November or December when there was a carload of Indians came up. They started to drive across the creek to the shack where we was at. Vic seen them coming. He had his deer rifle, and he took the rifle and went over to talk to these Indians. He knew them, well.

The thing of it was, this Indian's daughter was in the family way and she claimed Vic was the daddy. The old guy insisted that Vic marry the girl. Vic told the old guy he'd be down tomorrow. He left tomorrow, all right. He left, quit, and took off, and we never seen him again until January. He came back January 1st and said, "Just like coming home!" He told us he had heard that somebody else had married this Indian girl. She was a cousin of John Van Pelt, that Vic ran around with from Brookings.

Then I had one of those Alaska Indians work there, John De Heart. He was a pretty good man. He'd taken up with a White woman here in town. When I got him to go up, they was living in the apartment across from Koontz's Machine Shop. She wanted to go up with him. He said, "No, they don't have no place for you up there." She says, "What am I going to do? I haven't got any money." He says, "Go to the Salvation Army!" I felt sorry for her. She was a good looking woman. In fact, she was beautiful. John DeHart, he finally quit. He used to talk about how much money the government spent on educating him. He had pretty good education, John did. He liked to drink, but he died of the quick TB, not long after he quit.

George Baker

Then I had the old sawyer, George Baker. He was half Indian. He was a real good man, that fellow. I guess he weighed about 220 or 230 pounds, and stout as a bull. He knew his sawing job, too. I think he was 63 years old when he went to work for me. He worked 5 years. He was 68 when he finally quit.

Sometimes we couldn't get crew enough to run the woods and mill too. We'd go out and log, and get a bunch of logs in the pond. Then we'd saw them up. Old George, I'd let him punk whistle. He showed us plenty of stuff, old George did. He savvied how to do those things. I learned a lot of things from George. And those big logs. We had to roll them with a peavy. George Baker knew an old mill on the south end of town that had an overhead camming gear. So George said, "I think you can get that for any price you want to pay for it." So we went over and saw the guy. I got the equipment for $25 and we had to take it out of the mill.

He'd be sawing away, and get hold of a log where the saw wouldn't be cutting perfect enough. I'd tell him, "By God, George, we'd better take that saw off and get it pounded." "Naw," he'd say. He'd turn a little more water on the saw, back up, fill his pipe, and let her run. Puffing his pipe, he says, "She'll be all right when we get through with this log. This is a bad log." When he'd get through with that log, he might change his guide, to hold her into the wood a little bit more. Then when he got through with that log and got another, he'd set his guide back where it should be for an ordinary log. He didn't have much trouble.

Old Herman, down at the dock [in town], used to come over to the loads of lumber with his calipers and rule. He'd go down the boards like that. "How do you get'em so even?" he'd say. I'd say, "That old sawyer I got, and that farmer I got on the carriage." That's when Cle was setting ratchets.

Cle Wilkinson

I had to have five men or I couldn't operate. That's all there was to it. That's the least I could get by with. Once I got down to where I only had four men. I went to see if I could get Cle [Wilkinson] to come over. He said, "Now, you know I can't do that. What do you want me to do, anyhow?" I said, "I want you to set ratchets." He said, "No, I can't do that. Besides, I just bought a team of mules to pack ferns on. I gotta use them." I said, "Those Japs are sinking our ships right off the coast. If we don't all get our shoulders to the wheel and start rolling, those Japs is going to have control of this country." So, Cle came over to set ratchets.

Old George [the head-sawyer] you know, he liked to no more than stop and he was right back into the cut. But Cle, he wasn't that fast. George would come back and Cle would start jacking 'er out. George would start ahead and get-er into the saw, and Cle would just keep jacking. Push it right into the saw! George would have to back it out quick, and let Cle get it where he wanted it. It had to be on the mark before Clea would stop jacking.

Well, it didn't take George long to learn that he'd have to wait till Cle got it out where he wanted it. That worked pretty good. Finally, I told Cle, "You're the best ratchet setter we've ever had." They was even. George couldn't hurry him up.

This ends my dad's narrative, the oral history of his mill "above the falls." I thought that there was another tape, but it seems to be lost. The small sawmills such as Dad owned over the years had metamorphosed from the tie-mills of the 1920's and '30's to the small mills of the World War II era and finally they came into their own as the ubiquitous "Gyppo Mills" of the decade following the end of World War Two. The final section of this paper consists of my summary of the five mills Dad had from the end of the war until his last one, at Forestville, California, in 1956. That was the end of the era.

An Epilogue: How it all Ended

Before we got through there, we bought the ranch [at Hollow Stump, about 5 miles north of North Bend]. We had got a contract with Al Peirce for 1100 acres of timber. We had to go through this ranch to get it, so I bought these two ranches to get the right of way. We done pretty good on that setting.

I went to making ties there. George [Baker] was too slow for that, and he also didn't understand that you get the fine grain at the edge of the tie--the fine grain out in the sap. He'd make a tie up all right, but he wouldn't pay no attention to the sap. The top end is ahead on your log when you're cutting third growth and trying to get the grain. You pull it out about where you want it, then pull just the little slack in the rack, and you've got it. Just the bark is about all you're taking off of the edges of it. What few fine grains there is, you've got in your tie.

George didn't understand that, and I jumped in one day to show him. I fought the tie over, and showed him how to do that. He just stepped back. "This mill ain't going to be no good, anyhow," he said. That's what he thought. That mill made more money that all the rest of the mills I ever had! That's when he quit.

Hollow Stump mill under construction, 1945. Drag saw, left, log rollway, right. Roof is unfinished.

Inside the mill, under construction. Dad standing alongside of the GMC diesel power unit. Note the notched guyline stump inside the mill!

Dad standing in front of the GMC diesel power unit. He had two of those engines to run the mill – one to run the headrig and feedworks, the other ran the edger, trim saws, and conveyors.

Dad among the many pullies, belts and shafts of the mill.

The mill at Hollow Stump (1945-1948), and the End of the War

Dad's contract with Al Peirce called for him to produce a minimum of 10,000 board feet per day, and he paid Al Peirce a stumpage price of $2.00 per thousand. The war ended while he was sawing crossties in the new mill. If I remember correctly, he was getting almost exactly the same price for his ties there in 1945 as he had gotten in Cowlitz County Washington back when he first started in 1927, about $18 per thousand.

Lumber was a strategic material during World War II, and there were wage and price controls in effect for all lumber products, which explains why the price had not increased – and why inflation was kept in check even while wartime production was at maximum capacity. The role of the timber broker was almost eliminated during the war because the government would not pay the brokerage fee. All government orders for lumber were placed directly with a select group of large sawmills which invariably took larger orders than they were able to fill on their own. The remainder of the order – that beyond the ability of the large mill to produce – were subcontracted to the various small mills, giving the large mills a vested interest in keeping the small mills in business, and an unprecedented contol over many of them.

Grading specifications before and during the war were quite stringent and therefore the low-grade "third growth" Douglas fir was only marginally useful. The railroads demanded that the top one-inch of the cross-tie have at least six growth rings. The young, 65-year-old Douglas fir timber

79

near the bay often did not have six growth rings anywhere in the log, and if there was, it would be at the very outer edge of the log. It was a tricky manoeuver to be sure each tie had its required six grains in the top outer inch, as described above in Dad's description of his problem getting his sawyer, George Baker, to take enough care to get those six grains.

Those problems ended quite abruptly in November, 1946, when the Federal Government eliminated the wage and price controls that had been in effect during the war. Throughout 1947 and 1948 the price of lumber steadily increased at the same time that grading specifications were relaxed due to the high demand for lumber in the post-war housing boom. The demand for standing timber became acute and the price of stumpage went up – from $2.00 per thousand to $8, then $12 per thousand. Dad asked Al Peirce if he would be raising his stumpage. Al said, "no, we've got a contract." As a result, Dad made more money on that little sawmill at Hollow Stump than he made in all the other mills he had from 1927 until 1956. Having that wartime contract in effect at the time that wage and price controls were lifted was a lucky, once in a lifetime event.

The mill at Willanch, 1948-1951

By the time the Al Peirce contract was completed in 1948, timber for small mills had become hard to find. During the two years following the lifting of wage and price controls, almost every privately owned patch of timber suitable for a small mill had been secured by someone – speculators, sawmillers, or others. George Vaughan, son of early Coos County timberman William Vaughan, had scattered tracts of timber around the county and was busy setting up a mill on the West Fork of the Millicoma River. He needed a little more cash flow and he didn't need a 160 acre tract that he owned on Willanch Slough. Dad purchased it from him and built this, his fifth sawmill in Coos County in the timber there.

It was at the mill on Willanch Slough that I first worked in the woods. I was 15 years old and it was the summer after my freshman year at North Bend High School. My job was punking whistle, the job my mother sometimes performed while babysitting my brother and myself when we were 3 or 4 years old. The logging operation there was typical of the way that logging was done at many gyppo mills in the Douglas fir region at that time. It was a "two-donkey" show with the donkey at

80

the mill rigged to a North Bend Skyline system, swinging the logs from the yarder, which was about 1000 feet out from the mill. The yarder was set up on a highlead system, going out another 900 feet or so. I was punking whistle for the yarder. The crew consisted of the donkey punchers for the two donkeys, the hook tender Jack Olson, the bucker Oris Peterson, and myself as whistle punk. That's all it took to keep the mill in logs, and it was cutting about 15,000 board feet per day.

George Vaughn sawmill on private land adjacent to the Elliott State Forest, 1955. The green timber in the background is State land.

Sketch of a North Bend skyline system swinging logs from a yarder tree. The swing donkey and the sawmill would be off the picture, to the right. This is the system Dad used to have a "two-donkey show" and could log 160 acres of timber to the mill with two donkeys. From Willamette 1925 catalog.

By 1951 there were at least 75 small sawmills, now called "Gyppo" mills, operating in Coos County. To all appearances, Coos County was experiencing an economic boom. The stimulus provided to the local economy by the small mills was profound. There were at least four large machine shops operating on Front Street in Coos Bay, all of them with more work than they could handle, with broken parts coming in daily from small sawmills out in the woods. There were four large logging and sawmill supply companies, providing new and used equipment. The restaurants and taverns were doing land office business, as was all of the rest of the service and retail trade. One or two very large sawmills cutting the same amount of lumber that was produced by the aggregate of the small sawmills would not have given a fraction of economic boost provided by the many small mills. For one thing, the profits of the large mills went to absentee owners, or to anonymous stockholders. The profits from the small, locally owned mills remained in the area.

The down side of the boom was that the area was quickly running out of small tracts of privately owned timber. Owners of the tracts that were available were holding out for higher stumpage prices, which by 1951 was as high as $18 per thousand, up from $2.50 per thousand 5 years earlier. It would soon go to $25 per thousand. Wages and other costs were also up, but the price of lumber, and its demand, lagged. Small sawmills that could make it in that environment had to be very efficient indeed.

By the time Dad had finished cutting the Vaughan timber on Willanch Slough, it looked as though he may not find suitable timber for another mill. But there are always surprises in the lumber business. Vic Graham, the young man who had started working for Dad at McKinley back in 1939, had returned from the Navy following World War II and went back to work for him. Vic had come from a Seventh Day Adventist sawmilling family in Brookings, Oregon, and although he was not in any way an observing member of that community, he had connections. There was a patch of timber at Gravelford, on the North Fork of the Coquille River, and Vic made a deal for it with the widow of the owner. A mill can only exist if there is a timber supply, and Vic came through with the timber and talked Dad into taking him in as a partner and so was formed the Youst and Graham Lumber Company. That was the first (and only) partner my dad had taken since his first sawmill in

Washington, Youst and Rector, back in 1927. The Gravelford mill ran through 1951 and 1952 after which Vic went his own way with a sawmill near Bandon.

The Mill on the West Fork 1953-4

Dad found timber on the West Fork of the Millicoma River, near Allegany, and for the first time he encountered what was soon to be the bane of small operators. Regulations! When chosing a millsite, a prime consideration would be an accomodating canyon to fill with sawdust and slabs. The mill was already built with the conveyors and slab chutes heading out over the canyon when Jerry Phillips, a brand new employee of the Coos County office of the Oregon State Forestry Department arrived to inform my dad that he had to have a wigwam burner. He couldn't chute his slabs into the canyon, nor could he fill the canyon with the massive sawdust pile that would accumulate. Those days were over. Dad could not believe his ears. He had to turn the sawmill around, so that the waste went out the other direction and into a burner that he had to purchase and assemble. It was the beginning of the end for the independent small sawmill!

Dad's first wigwam burner being assembled at the Ash place near Allegany, Oregon, 1953.

There was another change. The old established sawmill people that Dad had been doing business with for the past 17 years were disappearing. To insure cash flow while he was setting up the mill on the West Fork, Jim Lyons (of Irwin Lyons Lumber Company) had advanced Dad $10,000, secured with nothing more than a simple promissary note. This would have been a standard business practice among sawmill men who had been doing business with each other for years. However, in February,1953, Jim Lyons was killed in a hunting accident. Some time later, two lawyers came to the house. They had found the promissary note lying in the top drawer of Jim Lyons' desk, and they told Dad, "We can't do business the way Mr. Lyons did. We'll need security

83

on that loan." The lawyers then drew up a three-page chattel mortage in which "one Coos-King donkey and two GMC diesel engines" were put up as collateral for the loan. This was obviously the end of an era.

The All Electric mill in California (1955-6): the last mill

The worst was yet to come. When Dad finished with the timber on the West Fork in 1954, my mother was becoming enthused about California. My sister was long married, my brother was in the Marines and I was in the Air Force – this was the Cold War and we all had to do our stint in the military. Dad had hurt his back and was in therapy in Texas. Mom was apparently quite bored when Raymund Schaecher of Schaecher-Kux Lumber Co., a partnership with interests in timberland in British Columbia, Oregon, and California, approached her on the subject of building a medium sized sawmill near some of their timberlands about 60 miles north of San Francisco. She went to the site of the proposed mill at Forestville, in Sonoma County, and fell in love with the romantic aspects of living near the Russian River and the Redwoods, and San Francisco, and the wine country. She was prepared to convince Dad that a sawmill right there would be the the thing to have. She was 48 and Dad was 55. Thus – without even looking at the Schaecher-Kux timber – they entered into an agreement that was the end of the Youst lumber business.

They subdivided part of the ranch, sold the house and a couple lots and the ranch itself to build a modern electric sawmill in a part of the country they had no knowledge of, or experience with. The site of the mill, at Forestville, was at the end of a spur railway line of the old inter-urban Santa Rosa and Petaluma Railroad about ten miles west of Highway 101. The Fibreboard Corporation had a large log storage deck and reload on the site, there was a mill pond from a previous sawmill, and it all looked very good.

Problems began to pile up from the beginning. The electrical equipment needed for an electric sawmill of the size he was building – 30,000 board foot per 8-hour capacity – was much more expensive than Dad had anticipated. He knew nothing about electricity and was continually surprised at the cost and complexity of it. He had no credit in California and where his credit was good – Coos Bay, Oregon – the bank was reluntant to make loans to an enterprise that would take place in another state. As a result, he became much too dependant upon credit from Schaecher-Kux Lumber Company. By the time the mill was completed he was in debt to them to a greater degree than was comfortable.

But the mill itself was a wonder to behold. Of the dozen sawmills he had built since his

George Youst sawmill, Forestville, California, 1955. This was an all-electric mill, with a planer, and 8-hour capacity of 30,000 bd. ft.

first one in 1927, this was the masterpiece. Fast and comparatively automated – it had live rolls, green chain, electric feedworks, and a planer mill to finish the lumber for the retail market. It was beyond the catagory of "gyppo." It had arrived! I remember overhearing a conversation in the Forestville Tavern about the construction of the mill. Referring to Dad as millright someone said with amazement, "He sure knows a lot about sawmills!" Indeed he did.

Truck and trailer load of redwood lumber leaving the yard of Dad's Forestville mill, heading for the California retail market, 1956.

Then the real problem exposed itself. The timber, some of which was coming from the old Jack London ranch at Valley of the Moon, was rotten. Mature Douglas fir growing that far south tends to have too much rot, and the timber that the Schaecher-Kux company had secured was so bad that some days a third of the wood that went through the mill had to be sent right on out to the burner. It was untenable. Then came the electric bills. In the place of a few hundred dollars a month for diesel, this mill took several thousand dollars a month for the Pacific Gas and Electric Company. Electricity had a certain convenience, but it had its price.

Compounding those problems, the lumber market in 1956 was in a severe recession. Prices were down, demand was down, and mills were closing throughout the west. Even Schaecher- Kux was having problems. Raymund Schaecher arrived at Santa Rosa with his accountant asking for immediate payment of an amount he said was owing him. Mom went to the books, pulled out the cancelled check, and showed Schaecher that it had been paid. Dad described Schaecher as turning white and clutching his chest. His accountant opened a bottle of nitro-glycerin pills and put one under his tongue. The economic recession was aggrevating angina throughout the west.

Mom knew better than Dad what the state of the business was, and it was bad. And, she shared an office with Harvey McGuire, the superintendent of the log reload of the Fibreboard Corporation. Suddenly and without warning, she and Harvey took off for Reno where they both divorced their respective spouses and married each other. I was 22 years old and had been around a little, but I was shocked to tears. Mom and Dad had been married 31 years. Welcome to California!

I cannot imagine the trauma that a divorce must inflict on younger children when I reflect upon my own reaction. I asked Dad what happened. He answered philosophically, "Sometimes women go crazy when they get to a certain age." If the divorce was a shock to me, it was devastating to Dad and he gave up on the mill. He arranged a lease-purchase to a California sawmill man who, in a few months, gave it back with $25,000 in unpaid electric bills and Federal payroll tax that he had witheld from the paychecks but had not paid the government. Pacific Power and Light and the Federal Government both came after my dad, with a vengance.

To clear those debts and avoid bankruptcy, Dad did a remarkable thing. He disassembled that entire sawmill by himself, piece by piece, over a period of about eight months. Some of the equipment weighed many tons. For power he had only the use of the jackscrew, the lever, the inclined plane, and the pulley. His back was never good after that, but in due course he cleared all the debts. There was no bankruptcy, which for sawmill men of his generation would have been an unacceptable disgrace.

Back working for wages, 1957-1961

Dad was 58 years old and would be going back to work for wages for the first time in 30 years. He told me that had been expecting to get another sawmill through the help of Al Peirce. Al was one of the most successful sawmill men of his time, having built the first "Swedish gang" sawmills on the Oregon coast, a sawmill specifically designed to cut the small "third growth" Douglas fir. He understood far ahead of his contemporaries that the future of the sawmill was in small logs, and capitalized on that understanding. Dad had a great respect for Al and had many business dealings with him over the years. Then, in May, 1958, Al Peirce bought a brand new Ford Thunderbird sports car in San Francisco and on the way home to Coos Bay ran into a redwood tree and was killed instantly. That was the final blow for Dad. The last of his solid connections to his sawmilling past had been severed.

He got a job scaling logs and running the log dump for Sweet Lumber Company, which had recently built a sawmill at Point Reyes, California. By a weird coincidence, Vic Graham – who had worked for Dad off and on since 1939 and was his partner in the Youst and Graham Lumber Company in 1951-2 – was the "pond monkey." Dad scaled the logs as they came in on the truck and then dumped the logs into the pond while Vic ran the Log Bronc, herding the logs across the pond to the mill. Meanwhile, Dad married Edna Peterson, a widow from Santa Rosa, and he commuted to Point Reyes from there. Vic lived at Point Reyes.

The Sweets were a prominent Coos County family and well known to Dad. They had purchased the timber at Point Reyes, about 30 miles from San Francisco. The timber was probably not much higher quality than the timber that helped to run Dad out of business, but the Sweets had an ace in the hole: environmentalists and conservationists were lobbying to have the timber on Point Reyes preserved as part of a proposed National Seaside Park. If the Sweet Lumber Company could play its cards right, there was a very good chance that the state or federal government might purchase it, and thus pull them out of a bad situation. That in fact is what eventually happened. In 1962, President Kennedy signed the law forming the Point Reyes National Seaside Park, and Sweet's logging road is now part of the 140 miles of wildernes trails within the park. The mill pond is the ecologically significant "Five Brook Pond," and there is no sign that a sawmill was ever there.

Dad's chance to get out of California came with a job scaling logs and running the log dump at a Schaecher-Kux plywood mill near Eugene, Oregon. He and Edna moved there, and he kept working until he was 62 and able to draw his Social Security. A very pedestrian end to a fascinating life that left a lot of sawdust in the woods of three west coast states.

Personal Addendum: 1957

My four-year enlistment with the Air Force ended on April 29, 1957. After my discharge, I visited Dad and Edna in Santa Rosa and headed on north, expecting to find work easily, as I had done before going into the Air Force in the first place. I stopped at Garberville, California where a cousin (Bud Huff) was working and I landed a job setting chokers for the newly formed Mathews Logging Co. Mathews' wife had inherited some Douglas fir timber which, for logging, required several miles of road to be built. I went to work with Mathews while he was still building the road, I helped rig the loading tree, and worked two weeks setting chokers behind his D-8 Cat. But, like the timber that put my dad out of business, the Douglas fir had too much rot in it. Mathews Logging Company went out of business after less than a month.

I drove north to Port Orford, Oregon, and went to work logging for Shaw Brothers Lumber Co., a gyppo sawmill. They had a nice operation, logging with a yarder and swinging the logs to the mill with a D-8 Cat and arch. I worked there a month exactly and they folded. The Shaw brothers were also broke, not because of bad timber but because of a bad economy.

Next, I went to work for Griffey and Laird Logging Company on Flores Creek near Langlois, Oregon. I worked there until late September when early rains washed out one of the bridges. I asked Floyd Griffey when they might start up again. He said they might not start again until Spring. This was really disappointing. At that same time I got a letter from the Air Force saying they would take me back with my old rank of Staff Sergeant. I took them up on it because the handwriting was on the wall. The glory days were over, when sawmills left their sawdust in the western woods. I needed a new line of work, and so did a lot of other people.

The End

Glossary

Donkey and Yarder: Donkey is the generic term for a stationary hoisting engine. A Yarder was a donkey mounted on a log sled and used for pulling logs from the woods to a landing. After about 1915, yarding was done by "high-lead" or by a "sky-line," in which lift was provided by a "spar-tree." These systems are still in use today. Steam donkeys were rated by the size of their pistons. Dad's 10 x 12 Seattle had a piston with a 10-inch bore and a 12-inch stroke and was built by Seattle Boiler Works. His 9 x 10 Willamette had a 9-inch bore and a 10-inch stroke and was built at Willamette Iron Works in Portland. By the mid 1940's most yarders were either diesel or gasoline instead of steam powered. Modern yarders are not called "donkey's and are no longer mounted on log sleds. They are mounted on rubber tires and have steel towers to provide lift.

Sawmill: This is the generic term for a mill that manufactures lumber from logs. A "cargo mill" was a mill set up on tide water,cutting lumber for shipment by sea. A "tiemill" was a small mill set up in the woods to cut railroad ties, usually averaging from 10 to 20,000 board feet per day. A small sawmill is generally considered one that produced less than 50,000 board feet of lumber per day. The post World War II boom brought about a proliferation of small sawmills and those cutting less than about 20,000 per day and set up in the woods came to be known as "Gyppo mills," and by extension, the tie-mill of a previouis generation came retrospectively to be called gyppo mills as well.

Parts of a Sawmill:

(a) *Headrig*, includes the head saws and the saw husk, mandril, belts and shafting. May also be called the timber sizer.
(b) *Carriage*, on which the log rides as it is fed through the headrig.
(c) *Feedworks*, controls the movement of the carriage, and is controlled by the Head Sawyer.
(d) *Setworks,* controls the lateral movement of the log for each succeeding cut. In the small mills this was done manually through a **ratchet** mechanism, operated by the "Ratchet Setter," who rode the carriage and responded to hand signals from the head sawyer.
(e) *Edger,* with its four to six saws, sawed the cants from the headrig into boards of varying widths.
(f) **Roll case**, the set of rollers on which cants, lumber,and slabs were moved through the mill.
(f) *Cutoff saws,* trimmed the lumber and cut it into specified lengths.
(g) *Conveyors,* removed the waste sawdust, slabs, and trimmings away from the mill.
(h) *Sorting tables,* used for sorting the lumber and piling it for movement away from the mill. The larger mills include a "**green chain**" which moves the lumber along under power.
(i) *Planer,* removes the rough surface left by the headrig and finishes the lumber into exact dimensions. Most of the small sawmills had no planer. They either shipped their lumber rough, or had it planed at a seperate planing mill.

Readings selected from the author's personal bookshelf

Ralph W. Andrews. ***This was Sawmilling***. Superior Publishing, Seattle, 1957. An outstanding collection of historic photographs of Northwest sawmills, large and small, with knowledgeable commentary.

Curt Beckham. ***Gyppo Logging Days***. Myrtle Point, OR. 1978. Memoir of a gyppo logger of the 1930's and 40's, with many colorful anecdotes.

Margaret Elley Felt. ***Gyppo Logger.*** University of Washington Press [1963], 2002. Expanded from her August 30, 1952, *Saturday Evening Post* article "I'm a Gyppo Logger's Wife." This highly intelligent and entertaining memior brings the gyppo experience to life. This new edition has an insightful forward by Robert E. Walls.

Grimshaw, Robert. ***Grimshaw on Saws***. Philidelphia, 1880. Reprint by Astragal Pess, Moristown, NJ. Concerns the history, manufacture, care and use of all types of saws. The classic on the subject.

Paul Hosmer. ***Now We're Loggin'.*** Binfords & Mort, Portland, 1930. A series of ironic articles each addressed to a specific logging or sawmilling specialty: The Millwright, The Sawmill Foreman, Sawyers and Setters, and so on.

Stewart H. Holbrook. ***Holy Old Mackinaw: A Natural History of the American Lumberjack***. The Macmillan Company, 1938, 1956. The classic account of the migration of loggers across the northern tier of states from New England to Oregon and Washington from the 1880's to the 1930's.

Ken Kesey. ***Sometimes a Great Notion***. Viking Press, 1964. Many reprints. This is the classic, timeless novel of the gyppo subculture in Oregon.

Joyce L. Kornbluh, ed. ***Rebel Voices: An I. W. W. Anthology***. University of Michigan Press, 1964, 1968. See especially pp. 282-285, "Johnson the Gypo," by Ralph Winstead. Reprint of an article which was among the very first in which the term appeared.

E. B. Mittelman. "The Gyppo System." In ***The Journal of Political Economy***, vol. 31, no. 6 [December, 1923]. The University of Chicago Press. An early scholarly attempt to describe the phenomenon. Very insightful.

William G. Robbins. ***Hard Times in Paradise: Coos Bay, Oregon, 1850-1986.*** University of Washington Press, 1988. Based largely on oral interviews, the life and values of the gyppo subculture come through clearly.

A. C. Samuel. ***Oregon Logger: Life and Times of A. C. Samuel.*** Maverick Publications, Bend, OR, 1991. Entertaining and folksy, it recreates the glory days of the 1920's and 30's.

The Timberman: An International Lumber Journal. Published in Portland, Oregon, from 1899 to 1955 by its founder, editor, and publisher, George M. Cromwall (1867-1950) and his son George F. Cromwall. The develpment of the small sawmill is documented quite clearly by articles and advertisements. Following are articles that are especially relevant to the subject of small sawmills:

"Small Sawmills," by G. A. Griswold. 7/28, p. 166ff.
"Future of the Demand for Cross Ties," by Earl Stimson. 7/28, p. 170ff.
"What is the Future of the Wood Cross Tie?" by G. C. Warne. 7/30, p. 162ff.
"Tie Making at Morton: Interesting Small Mill Center." 9/30. p 156.
"Requisites of the Successful Small Mill," by John S. Webster. 4/31, p 58ff.
"The Small Sawmill Industry in the Douglas Fir Region," by Herman M. Johnson. 10/33, p 9 ff.
"Diesel Power Unit in a Portable Sawmill." 2/35, p. 24.
"The Small Movable Mill," by Paul Lueth. 7/37, p 5ff.
"The Small Sawmill in the Douglas Fir Region," by Herman M. Johnson. 12/44, p. 34 ff.
"Portable Sawmills Become Precision Machines," by Terry Mitchell, 03/45, p. 47.

Timber Times. A journal of logging and lumbering history and modeling, published quarterly at Box 219, Hillsboro, OR. The 45 issues published between 1993 and the present contain a treasure trove of historic logging and sawmilling photographs and many significant articles on the subject.

Useful Facts: Sawmills and Saws. A 47 page pamphlet compiled by R. Hoe and Company, America's oldest saw manufacturer, with useful infomoration relative to the small sawmill. Third edition, 1947

Robert E. Walls. "Lady Loggers and Gyppo Wives: Women and Northwest Logging." **Oregon Historical Quarterly** 103 (Fall, 2002). Groundbreaking work on the role of the wives of gyppos.

C. H. Wendel. ***The Circular Sawmill***. Stemgas Publishing Co, Lancaster, PA. 1989. This book contains 67 pages of reprints of historic sawmill advertising which provide good pictures of the workings of small circular sawmills typical of those used during the first half of the twentieth century.

Lionel Youst. ***Above the Falls: An oral and Folk history of Upper Glenn Creek, Coos County, Oregon.*** Coos Bay, OR [1992], 2nd Revised Edition, 2003. See especially Chapter 7, "The Timber," which includes a transcription of George Youst's oral account of his gyppo sawmill in Coos County before and during World War II.

Index

A
Aason John 50-1
accident, fatal 52-3
accident, truck 69, 70
Acme Coat Hangar Company 4, 24
Alaska Junk Company 39
Allegany, OR 44 map, 62, 74, 83
American Sawmill Machinery Co. 16
Arcata, CA 50
Atwater=Kent radio 30
Arcata, California 50

B
Baker, George 76, 78
Bandon, OR 48
Benham and Laird Logging Co. 58
Benham, Julius 58
Bingham, William (Geo Youst's uncle) 15
bridges 30-31, 37, 70
bridge plank 5, 30
Bridges, Harry 35-6
broker 5,6 18, 25, 46, 61
Browning, Dave
Brookimgs, OR 59
bull block 33
Burke, Bill 17, 24
Burr, Robert 42
Busterud, Herbert 5, 45-6, 49, 54, 61-2
Button, C. A. 24

C
carriage 31, 62
Centralia WA 3, 13, 15, 27
chainsaw 7-8
Chehalis WA 3,13,14, 21
Chevrolet 15
chute 64-5
conveyor 6
Coos Bay Iron Works 48
Coos Bay Lumber Company 5, 55
Coos County, OR 46
Cooston 44 map,
Coquille River 54, 82
Cowlitz County WA 4, 12 (map), 79
Cross ties 4, 6, 7, 13, 16 definition, 23, 36
cutoff saws 6

D
Detroit Diesel 6

Dillon, L. L. 24
dimensioned lumber 7
Dodge "fast 4" 39
Donahue, Frank (independent contract logger) 8
donkey sleds 48, 62, 64
donkey, steam 31-4, 38, 48, 55-7
donkey, gas 48, 54, 56, 66, 80-1, 84
Douglas fir 5, 46, old growth 53, third growth 79, 87
Dugger, Dave 55
Dynamite caps 47, 49

E
Eastern Lumber Co. 14
edger 6, 21
Eugene, OR 88
Eureka, CA 51

F
Fairhurst Lumber Co. 5, 19, 24, 25, 45
feedworks 15, 17, 62-3
Fibreboard Corporation 84, 86
fire 6, 20, 45, 48, 49, 53, 54, 59
Ford 15, 27-8, 48, 87
Forestville, CA 84-6
Francis, Doc 14-15
friction blocks (logging donkey) 48
friction feed (sawmill) 17, 62

G
Galvin, WA 14
Garberville, CA 88
General Motors 6
grading specifications 21, 54, 79-80
Graham, Vic 56, 59, 82
Gravelford OR 44 map,, 82
Griffey and Laird Logging Co. 88
Gunderson Brothers 6
guylines 33, 52, 53

H
haulback 32, 50
Hauser, OR 42, 44 map,
Hearst newspapers 36
H. H. Martin Lumber Co. 14
Holbrock, Stewart H. (author) 8
harness 19, 25, 28
horse logging 18-20, 25-8
Huff, Bud 42, 49, 51, 54, 58-9, 72
husk 15, 17

I

Indians 47, 76
internal combusiont engines 5, 22, 23, 24
Irwin-Lyons Lumber Co. 49, 54
IWW 8

J

Jack London Ranch 85
Johnson, Gus 74

K

Kalama, WA 4, 14, 18
Kelso, WA 34
Kesey, Ken (novelist) 3

L

LaCenter WA 38
Landers, Kenny 66, 68-70
Landers, Lionel 66, 68
Levison, Dave 54-60
Lewis River 31
Linden, John 42
Little Kalama 4, 13 24, 37, 42
longshormen strike 1934 35-6
Longview WA 35, 41
lumber market 5, 7, 9, 24. 28, 80, 82, 86
Lusignot, Pete 74
Lyons, Jim 49, 51, 83-4

M

mainline 32
Mathews Logging Co. 88
McGuire, Harvey 86
McKinnley OR 44 map, 54-60,, 82
Melwin Dam (Lewis River) 41
mine timbers 29-30
Millicoma River 83
Moore Mill and Lumber Co. 54
Morton, WA 19
mules 47

N

neighborhood mills 6
North Bend, OR 5, 78
North Inlet 52

O

Ogle, Dude 23, 35, 42, 66, 67-8
Olson, Jack 81
Olympia, WA 13
Oregonian 24
OWR&M RR 24

P

Peirce, Al 79, 80, 87
Peterson, Ernest 46, 49
Philips, Jerry 83
plank road 29, 73
planer 7, 85
Point Reyes, CA 87-8
portable sawmills 38
Portland Cow and Horse Exchange 19
Port Orford, OR 88
power lines 41
Powers, Tom 42

R

Rector, Earl (Curly) 15, 19, 21-2

S

sawmill, elecric 84-6
sawmill, portable 37-8
sawmill, Swedish gang 87
Schaecher-Kux Lumber Co. 84-6, 88
Schapers, Joe 61
Schnitzer, Sam 39
Seaburg (fatal accident) 53
Security State Bank, Woodland, WA 24
Seventh Day Adventist 38, 59, 82
Sharkey, John P. 28
Shaw Brothers Lumber Co. 88
Siestreem, Jim 46
Siuslaw Lumber Co. 48
Skinner, Louie 36-7
slabs 31
slab pile 20-1
spar tree 32-3, 40, 49, 72
splash dams 50, 71-4
steam 5, 23
Stevenson, WA 37, 42
strawline 32
Strickland, Roy 67
strike, 1934 longshorman's 35-6
Studebaker 38
Sweet Lumber Company 87-8

T

Tacoma, WA 24
Tie-mills 3, 4, 6, 13,14,18, 78
tightline 55, 58, 73
timbers 7

U

Union Mills 35
Union Steel and Rail Co. 32

V
Vancouver, WA 4, 24
Vaughan, George 80, 82

W
Washington Iron Works 32
water-power 5, 23, 63
Wesfall, Hank 40
whistle punk 53, 76
Wilkinson, Bob 60-2
Wilkinson, Cleland 60, 73, 77
Woodland WA 4, 12 map, 14, 24
World War I 3
World War II 3, 9, 77, 79
WPA 36-7

Y
yarder, see donkey
Yelm, WA 15, 17
Yost, Bill 34-5, 39
Yost, Swede 35
Youst and Graham Lumber Co. 82
Youst, Doris 8, 53, 68, 84-6
Youst, George (interviews with) 3, 13
Youst, Laurence 47, 84
Youst, Lionel 53, 84, 88

www.ingramcontent.com/pod-product-compliance
Lightning Source LLC
Chambersburg PA
CBHW051227200326
41519CB00025B/7272